Applications of Polymers in Surgery

Edited by

Inamuddin[1], Tariq Altalhi[2], Mohammad Luqman[3], Hamida-Tun-Nisa Chisti[4]

[1]Department of Applied Chemistry, Zakir Husain College of Engineering and Technology, Faculty of Engineering and Technology, Aligarh Muslim University, Aligarh-202002, India

[2]Department of Chemistry, College of Science, Taif University, 21944 Taif, Saudi Arabia

[3]Department of Chemical Engineering, College of Engineering, Taibah University, Yanbu, Saudi Arabia

[4]Department of Chemistry, National Institute of Technology, Srinagar, Jammu and Kashmir 190006, India

Published by **Materials Research Forum LLC**
Millersville, PA 17551, USA

Published as part of the book series
Materials Research Foundations
Volume 123 (2022)
ISSN 2471-8890 (Print)
ISSN 2471-8904 (Online)

Print ISBN 978-1-64490-188-5
eBook ISBN 978-1-64490-189-2

Distributed worldwide by

Materials Research Forum LLC
105 Springdale Lane
Millersville, PA 17551
USA
https://www.mrforum.com

Manufactured in the United States of America
10 9 8 7 6 5 4 3 2 1

Table of Contents

Preface

Utilization of polymers as biomaterials is rising rapidly day by day in medical surgeries. A biopolymer material should be easy to fabricate, production of cheap, lightweight, durability, biodegradation, ability to tailor mechanical properties, and should be easy to sterilize without alteration in properties. These polymer biomaterials are the materials that can be implanted in the body to provide a special prosthetic function or used as surgical, diagnostics and therapeutic applications. Tissue injuries hit purposefully or accidentally. A polymer biomaterial when it comes in contact with tissues and blood should not cause any harm or destroy components of tissues and blood. Further, it should not be toxic and allergic. Polymers play an important role in the medicine field and have been remarkably promoted to surgical applications for healing those injuries.

Applications of Polymers in Surgery provides a comprehensive overview of polymers for surgical applications such as synthetic blood substitutes, skin, joints, bones, retinal, dentures, cardiovascular, neuro and artificial internal organs, etc. Also included are fabrication, characterization, properties, known biodegradability for various types of polymer biomaterials in medical implants, surgical devices and supports. This book is envisioned for postgraduates, faculty, educators, research scientists and doctors in the fields of medicine, chemistry and biology, advanced material science and polymer chemistry. The work reported in the following 8 chapters are as follows:

Chapter 1 describes common polymers used in medicine, considering that adapting polymers for medical purposes involves varying parameters such as crystallinity and crosslinking degree. Examples of medical polymeric materials through selected physico-chemical properties and advances of polymers for improving health at the stages of diagnosis, monitoring and treatment are discussed.

The chapter 2 focuses on some advanced polymers like resin, copolymers and thermoplastics which are therefore used in different fields like medical or surgical in large scale on the basis of structure activity relationship. Application of advanced polymers and biodegradable polymers were also discussed.

Chapter 3 presents the numerous types of polymer biomaterial scaffolds for tissue engineering, different types of clinical applications and fabrication techniques for fabrication of the 3D scaffolds. A focus of hydrogel as scaffolds is discussed in detail. Meanwhile, the common interaction reaction between biomaterials and cells are also reviewed. The discussed information in this chapter is essential to be considered in any development of a scaffold as an implant for tissue engineering.

Chapter 4 details natural and synthetic polymers and their use in the medicinal fields. Owing to their properties, they are considered more suitable for neurosurgery compared to conventional interventions. Moreover, their tailored formulations, co-polymerization, degree of crystallization and blending with other polymers allow them to be applied in different neurosurgical procedures.

Chapter 5 focuses on the diverse significant roles of polymers for designing ocular systems including contact lens, artificial cornea, orbital ball, lacrimal duct, glaucoma drainage implants etc., and their strategies to combat ophthalmic pathologies.

Chapter 6 explores the advancement of guided tissue regeneration membranes in periodontal surgical techniques. Moreover, both non-resorbable and resorbable barrier membranes are discussed to acquire appropriate guidance for periodontal regenerative treatment. The composite membranes are also described in addition to the benefits and challenges of advanced barrier membranes over conventional membranes.

Chapter 7 discusses the different types of denture liners, their clinical application, properties, and recent advances in these denture liners. In addition drawbacks (surface roughness, adhesion, color stability, water sorption, and difficulty in cleaning) of lining materials have been highlighted. Some of the commercial products are also mentioned.

Chapter 8 details several polymers used in non-resorbable and absorbable membranes for guided bone regeneration. Additionally, the role of their physicochemical characteristics, advantages, disadvantages are discussed in detail. The focus is on advances which enable its characteristics for handling during surgery and act as a physical barrier, promoting an environment to regenerate bone tissue in a healthy way.

Highlights

- Includes an overview of the use of polymers in surgery
- Explains the utilization of polymers in hospitals
- Presents the latest innovations in this field
- Provides cost-effective solutions to many types of surgeries

Applications of Polymers in Surgery
Materials Research Foundations **123** (2022) 1-29

Materials Research Forum LLC
https://doi.org/10.21741/9781644901892-1

Chapter 1

Polymers in Clinical Medicine

R.E. Padilla-Hernández[1,a], A.L. Ramos-Jacques[2,b], A.R. Hernandez-Martinez[3,c,*]

[1] PCeIM-UNAM, Centro de Física Aplicada y Tecnología Avanzada (CFATA), Universidad Nacional Autónoma de México (UNAM), Blvd. Juriquilla 3000, Querétaro, México

[2] EditAcademy, México, 76269 Querétaro, México

[3] Centro de Física Aplicada y Tecnología Avanzada (CFATA), Universidad Nacional Autónoma de México (UNAM), Blvd. Juriquilla 3000, Querétaro, México

[a]ricardeph@gmail.com, [b]al.ramos.jacques@gmail.com, [c]angel.ramon.hernandez@gmail.com*

Abstract

The use of polymeric materials for clinical purposes has advanced through improved methods of synthesis and chemical modification. Adapting polymers for medical purposes involves varying parameters such as molecular weight, crystallinity, and crosslinking degree. Polymers for device development in medicine must meet specific characteristics (flexibility, ease of use, biocompatibility) for their use in diagnosis, monitoring, and treatment. Polymers can be used as antibacterial coatings, dialysis membranes and wound dressing applications, but polymeric materials in the field of clinical medicine have countless uses. This chapter describes an overview in three content sections: (1) summary of the most common polymers with medical purposes, (2) examples of medical polymeric materials through selected physico-chemical properties and (3) advances of polymers for improving health at the stages of diagnosis, monitoring, and treatment.

Keywords

Poly(lactic acid), Chitosan, PEG, Point-of-Care, Drug Delivery

Contents

Materials Research Forum LLC
https://doi.org/10.21741/9781644901892-1

1.   Introduction

The use of materials for medical purposes dates back to early human history, as has been found in ancient medical documents. For instance, Edgar Smith papyrus (18th Dynasty of Egypt around 4000 BC) included anatomical descriptions and wound treatments used in war, such as sutures and wound closure devices that are documented in hieratic writing [1]. Materials for clinical medicine have been refined over time, which is reflected in the differences between the rudimentary ancient materials and the latest and most sophisticated materials. For example, in bone fracture treatments, the materials employed for immobilization were wood, leather, bamboo and linen [2,3]; in contrast with using metals or plaster, and splints produced with synthetic polymers in 3D printers [4].

In the mid-twentieth century the polymer industry was a very important sector, and clinical medicine was one of the areas with more impact, due, among other factors, to the arising scarcity of conventional materials resulting from the events occurred during the Second World War that lead industries to search for alternative materials [5]. Today, clinical medicine continues using polymeric materials in the development of novel and better devices. This chapter describes polymeric materials in three content sections; (1) summary of the most common polymers with medical purposes, (2) examples of medical polymeric materials through selected physico-chemical properties and (3) advances of polymers for improving health at the stages of diagnosis, monitoring and treatment.

2. Polymers with medical purposes

Countless polymers are reported in scientific literature and used in the industry, due to the wide variety of chemical structures that can be obtained. However, some groups of polymers tend to be more commonly synthesized for reasons such as costs and ease of manufacture. Fig. 1, shows the most commonly used polymers for the medicine area.

Table 1. Medical uses of polymers. Modified from: Maitz, M.F. 2015 [6]

Uses	References
Ocular and Intraocular lenses	[7,8,9]
Organic sensors	[10,11]
Dental resins and prostheses	[12,13]
Instruments and components coatings	[14,15,16]
Heart valves and implants	[17,18,19]
Contrast agents and bioimaging	[20,21]
Hemodialysis membranes	[22,23]
Wound dressing	[24,25,26]
Brain implants	[27,28]
Tissue adhesives	[29,30]
Bone implants and healing	[31]
Catheters	[32,33]
Tissue engineering	[34,35]
Containers for sample collection	[36,37]
Suture threads	[38,39]
Vascular stents and grafts	[40,41]

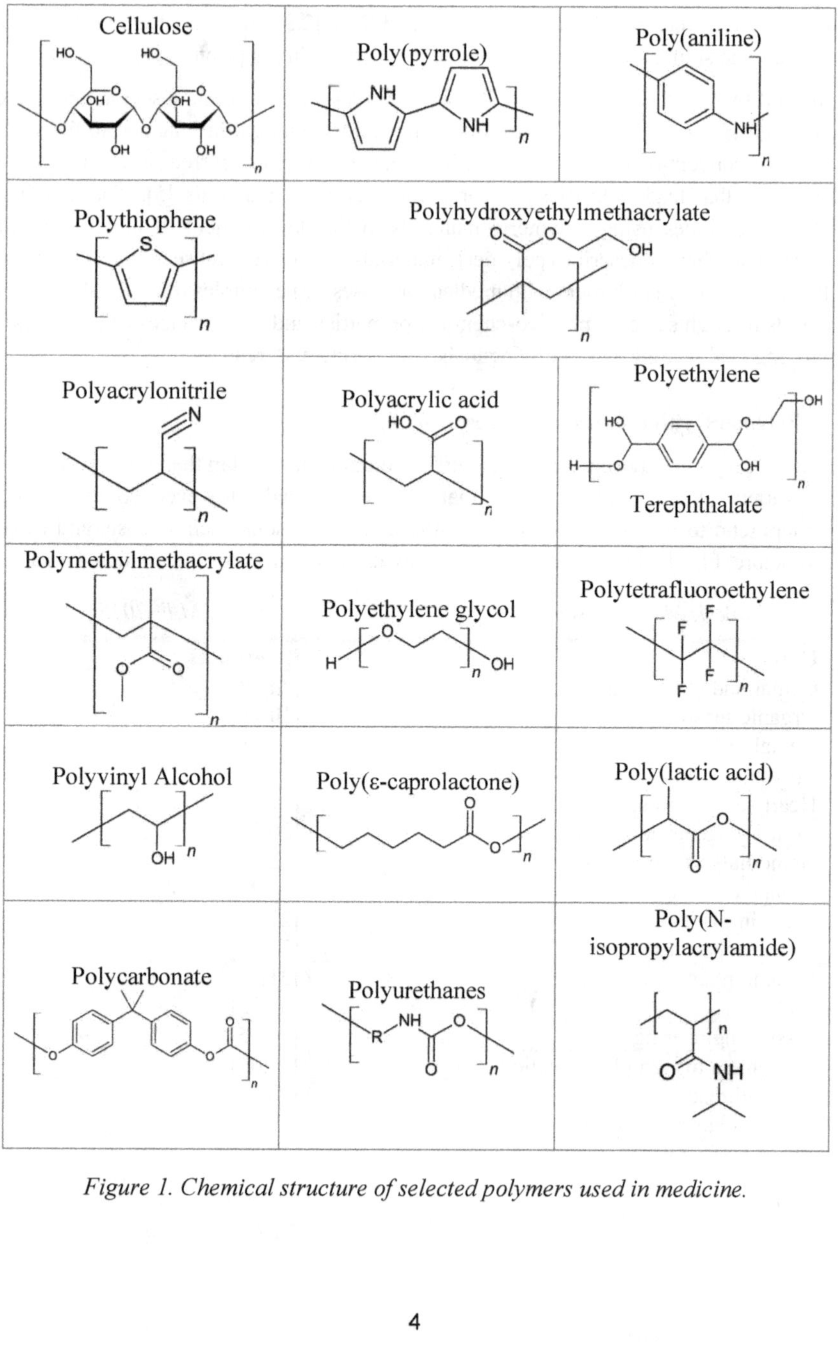

Figure 1. Chemical structure of selected polymers used in medicine.

Materials Research Forum LLC
https://doi.org/10.21741/9781644901892-1

For the development and research of materials, those polymers are synthesized, functionalized, or combined to provide particular physico-chemical properties for each use in the broad area of medicine. Some examples for solving specific problems in medicine are shown in Table 1.

3. Useful polymers characteristics

Specific characteristics of polymers such as origin and physico-chemical properties involve a particular interest for technological applications. This section describes examples of polymers in clinical medicine that take advantage of a specific feature.

3.1 Origin

According to their origin, polymers can be grouped into two general categories: polymers of natural origin (or biopolymers) and synthetic polymers. This classification is debatable as other authors have included the term "synthetic biopolymers", which are polymers of natural origin that have been chemically modified or polymers synthesized (from synthetic monomers) that do not generate toxic residues when they are degraded [42]. However, this major division will be used for explaining selected cases.

3.1.1 Natural

IUPAC [43] defines the biopolymers as macromolecules present in living things, including polysaccharides, polynucleotides and proteins. Polysaccharides are generally more stable under environmental conditions and are easy to process; then they are commonly used for several applications [44]. Within the medical field, polysaccharides are preferred due to their biocompatibility and low toxicity.

Several efforts have been made to develop technologies that allow biopolymers to be processed or isolated in order to produce materials with well-defined properties. For example, Pinho and collaborators [45] developed a textile system based on cotton cellulose functionalized with hydrogels to improve materials wettability and drug-releasing capacity. That work suggests that new functionalities can be incorporated into dressings for wounds such as antimicrobial properties or thermo-/pH-responsive sensitivity.

Mertaniemi et al. [46] developed extrusion suture threads based on cellulose hydrogels modified with glutaraldehyde as a crosslinker; in order to offer an alternative solution to the low retention of human adipose mesenchymal stem cells (hASC). In their study, they showed that these fibers increased their mechanical strength with the crosslinker even in humid conditions for wound treatment.

3.1.2 Synthetic

Synthetic polymers are those polymers obtained by chemical synthesis (not naturally produced). There are various synthesis routes used to obtain polymers, but mainly they are obtained by addition and by condensation reactions, considering the nature of precursor monomers[47]. Polymerization could be performed by thermochemical, electrochemical, and photochemical reactions, using catalysts or by the instantaneous reaction of the precursor species. Some examples of synthetic polymers are polytetrafluoroethylene, polycarbonate, poly(ε-caprolactone) (PCL), polymethylmethacrylate, polythiophene and polyurethane [48,49]. Movahedi et al. [50] manufactured and characterized structured core-layer nanofibers with the coaxial electrospinning technique of polyurethane/starch and polyurethane/hyaluronic acid for the treatment of wounds and skin tissue engineering.

3.2   Physico-chemical properties

Polymers for clinical medicine are subjected to rigorous testing stages; in particular toxicology, compatibility, reactivity (or innocuousness), and capability of preserving material properties. Therefore, it is essential to consider physical and chemical properties of polymers (Table 2.), for the correct selection of the material. Polymers can have different characteristics due to their complex and varied structures, and because they can be easily modified.

Table 2. Characteristic properties of polymers

Physical	Chemical	Biological
Mechanical	pH	Biocompatibility
Optical	Reactivity	Toxicity
Electronic and/or ionic	Oxidation/reduction	Biodegradability
Thermal	Combustibility	Pharmacokinetics
Hydrophobicity/hydrophilicity		

3.2.1 Mechanical properties

Mechanical properties are fundamental in applications that require materials to be subjected to tension, friction, and compression. Polymers exhibit lower hardness than ceramic or metallic materials. In clinical medicine, mechanical properties are important for applications such as textiles, implants, suture threads, coatings, among others.

Bittner et al. [51] studied PCL porous scaffolds and PCL-hydroxyapatite with incorporated uniform vertical porosity and ceramic content gradients, for evaluating the effect of porosity in printed materials. They use a 3D-printer multi material extrusion system, varying the weight percentage ratio and fiber spacing. Their results of the mechanical tests

Materials Research Forum LLC
https://doi.org/10.21741/9781644901892-1

showed compression moduli within the adequate values for trabecular bone scaffolds. On the other hand, Mondal et al. [52] focused on mechanical characterization and optimization of design parameters of poly(lactic acid) scaffolds with nano-hydroxyapatite obtained by 3D printing.

However, the importance of the mechanical performance of scaffolds for tissue engineering is not limited to materials intended for the bone system. Vogt et al. [53] obtained scaffolds by electrospinning solutions of sebacate aimed at engineering cardiac tissues. And they conclude that the mechanical properties of their material were higher in magnitude than native human myocardial tissue, which in the authors opinion may be beneficial in reducing infarct expansion and in remodelling the left ventricle.

Li et al. [54] conducted a study aimed at the development of braided suture threads from oxidized regenerated cellulose, because it is biocompatible, biodegradable and has homeostatic features. The data showed material tensile strength values above 10.0 N for all sutures, that are higher than those indicated by the United States Pharmacopeia (USP). There are also numerous studies on vascular stents, such as the one carried out by Hua et al. [55], whose work is focused on the study of the mechanical properties of two groups of fine-tip stents made from the laser cutting of poly(L-lactic acid) tubes. They performed three different mechanical tests under the ASTM F3067-14 standards: dog bone traction tests, radial load tests of tubes and radial load tests of stents. The aim was to determine the importance of knowing the radial support capacity of poly(L-lactic acid)-based stents.

3.2.2 Optical behaviour

Clinical medicine requires extensive use of optical devices for the correct functioning of diagnostic or treatment tools. Some recent studies show that those needs can be fulfilled by polymers. Fu et al. [27] focused on obtaining implantable and biodegradable optical fibers based on poly(L-lactic acid) (PLLA), as an alternative to silica fibers. The fibers are used in photonic systems for biomedical uses, such as laser surgery and phototherapy. PLLA fibers had high mechanical flexibility and optical transparency. The same fibers were implemented as a tool for intracranial light delivery and detection, performing deep brain fluorescence detection and optogenetic interrogation in vivo.

Neumann et al. [56] studied borylated poly(fluorene-benzothiadiazole) (PF8-BT) derivatives encapsulated in pegylated poly(lactic-coglycolic acid) nanoparticles and it was compared with a reference near infrared-emitting conductive polymer or Indocyanine Green . Neumann's work is supported by the hypothesis that PF8-BT are π-conjugated polymers with emission and absorption profiles of deep red / near infrared suitable for the formation of optical imaging in vivo. In their work they conclude that fully PF8-BT, sample P4, had a better performance than the reference conductive polymer as a photoluminescent and studied pH-sensitive hydrogel nanoparticles aimed at the controlled release of anticancer drugs. Polymerization was performed by reversible addition/fragmentation chain transfer, using poly (ethylene glycol)-diacrylate as cross-linking agent for obtaining the copolymer poly(hydroxyethyl methacrylate-co-N,N-dimethylaminoethyl methacrylate). Nanoparticles of 180 nm (pH 7.4) were synthesized with the swelling capacity in acid medium (pH 5.5) reaching sizes of up to 400 nm. Results showed a doxorubicin release of c.a. 20 % of the loaded drug in a 7.4 pH medium and 80% at a pH of 5.5. In a previous work, also Qin et al. [66] developed mixed poly zwitterionic-charged micelle solutions (aminoethyl methacrylate)-co-poly(methacrylic acid)-co-poly(n-butyl methacrylate); with the objective of been a drug nanocarrier for intracellular administration of doxorubicin. Micelles have a uniform and small diameter under physiological conditions; but in acidic conditions (pH 6.5), micellar aggregates are formed. They observed improved cellular uptake of micelles in acid media, and that behavior can favor micelles retention in tumor tissues.

4. Advances in polymer materials for clinical medicine by application

Clinical medicine is an area of medicine that involves direct dealing with patients, which involves diagnosis, monitoring, and treatment (Fig. 2.). This section discusses the more recent advances on polymers aimed to be used in clinical medicine applications.

4.1 Prevention and diagnosis

Disease prevention relies on the availability of diagnostic tools that could be affordable, effective, efficient and portable. Medical monitoring of health conditions provides essential data for detecting irregularities that can indicate a pathology. Metabolic activities could be daily monitored as the glucose that nowadays requires puncturing the skin to collect blood. The pain and discomfort associated with the use of that tool supposes a new need for developing non-invasive portable glucose devices [67].

4.1.1 Point-of-care testing

The point-of-care testing refers to obtaining patient health measurable indicators in the site, where caring is being carried out, with the immediate availability of the results. Based on this idea, many research works have focused on the study and development of devices capable of fulfilling this task. As seen in Fig. 2, sensors are one of the more broadly studied devices, including organic sensors based on conductive polymers (Table 3.).

Materials Research Forum LLC
https://doi.org/10.21741/9781644901892-1

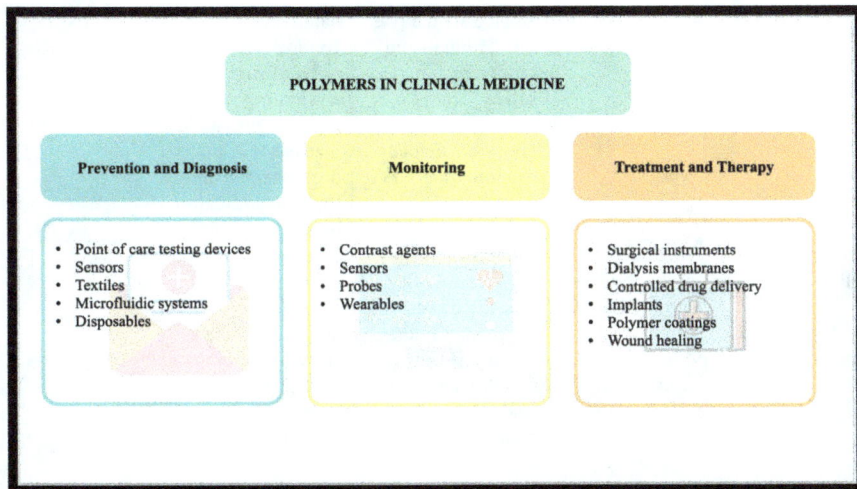

Figure 2. Applications of polymers in different areas of clinical medicine

Table 3. Polymers proposed for organic sensors development

Polymer	Additional materials	Sensor	Results/ Conclusion	Authors	Ref
Polypyrrole (PPy)	Multi-walled carbon nanotubes in Nafion, glucose oxidase and platinum electrode	*Glucose biosensor electrode* *high sensitivity of 54.2 µAmM −1 cm −2 *good selectivity	Oxidized PPy ensures a large number of active sites, improves biosensor stability	Shrestha et al.	[68]
Polypyrrole (PPy)	Epirubicin (anticancer drug), nitrogen-doped reduced graphene, salmon sperm ds-DNA, guanine	*Tag-free DNA-based biosensor* *detection of epirubicin (low concentration) in real-life samples	Guanine detection based on epirubicin concentration (0.004 - 55.0 µM)	Khodadad i et al.	[69]
Poly (carboxy-betaine methacrylate), polyCBMA	Carcino-embryonic antigen, CEA, (biomarker)	*Electrochemical sensor*	Material characterized with single protein solutions. PANi	Wang and Hui	[70]

Polyaniline (PANi) nanowire		*tested with cow's milk, human and bovine serum,, saliva *possible detection of biomarker CEA	nanowires conductivity with polyCBMA antifouling capacity provide a matrix for ultrasensitive and low contamination biosensor growth.		
Polyaniline (PANi)	Nickel, DNA probe, bovine serum,albumin ,	*Composite metal-organic sensor* *for the direct detection of non-amplified Hepatitis C (HCV) nucleic acid *glassy carbon electrode	Biosensor achieved quantitative detection of the HCV target in the presence of nonspecific nucleic acids	Sheta et al.	[71]
Polyaniline (PANi) Poly (4-vinylaniline) (PVAN)	Bacterial cellulose membranes	*Biosensor for neural stem cell differentiation*	Voltage-sensitive nanocomposites can interact with isolated neural stem cells from the subventricular zone of the brain without visible signs of cytotoxic effects	Rebelo et al.	[72]
Polyaniline (PANi)	Au nanoparticlesc arbon cloth,	*Glucose biosensor* *flexible non-enzymatic *favors glucose oxidation	The glucose sensor showed a linear range of 10.26 µM to 10.0 mM, a sensitivity of 150 µA cm^{-2} mM^{-1}	Xu et al.	[73]
Poly (3,4 ethylenedioxyt hiophene) (PEDOT) Polystyrene sulfonate (PSS)	enzyme glucose oxidase	*Glucose biosensor* *uses saliva * made on recyclable disposable paper substrate	Device showed sensitivity for detecting glucose in saliva and detecting abnormal concentrations	Bihar et al.	[67]

		*fully inkjet-printed and metal-free			
Poly (3,4 ethylenedioxythiophene) (PEDOT)	Graphene oxide, Indium tin oxide (ITO), uric acid, human saliva	*Paper sensor* *uses saliva *uric acid detection	The linear range for UA detection was 2 to 1000 μM, that corresponds to the detectable concentration in human saliva	Huang et al.	[74]
Poly[3-(3-azidopropoxy)thiophene]	Lactose or ferrocene, Cu (I), Erythrina crista-galli	*Electrochemical sensor* *developed in real-time *carbohydrate-protein recognition	The use of ionic liquids for electrochemical polymerization improves conductive properties.	Qu et al.	[75]
Poly (3,4 ethylenedioxythiophene) (PEDOT)	Au nanoparticles anti-vascular endothelial growth factor (VEGF) antibodies	*Electrochemical sensor* *detection of VEGF	Linear relationship with VEGF concentration (1-20 pg/mL), detection limit of 0.5 pg/mL	Kim et al.	[76]
Poly (3,4 ethylenedioxythiophene) (PEDOT) Polyglycerol (HPG)	2,3-dihydro thieno[3,4-b][1,4]dioxin-2-yl)methanol, alpha-fetoprotein antibodies	*Electrochemical sensor* *complex biological media	Sensor reached a detection level of 0.035 pg/mL in a complex biological sample	Ma et al.	[77]

Despite organic conductive sensors, there are also other sensors that are based on different principles. Faalnouri et al. [78] studied nanosensors based on surface plasmon resonance (SRP) that are made from polymeric films and nanoparticles of poly(hydroxyethyl methacrylate-N-methacryloyl-(L)-glutamic acid). They used ultraviolet polymerization for the detection of amoxicillin in commercial and local milk samples using the molecular imprinting technique. According to their results, the linearity range was measured as 0.1-200 ng/mL for amoxicillin-printed polymeric film and nanoparticle-based SPR nanosensors.

4.2   Monitoring

Although monitoring in clinical medicine can also be associated with both diagnosis and treatment, this section mainly exemplifies most recent studies on those materials for the development of devices for constant review of biological signs.

4.2.1 Imaging monitoring

Riaz et al. [79] developed a poly-2-hydroxyethyl methacrylate (PHEMA) contact lens with an anthocyanin dye extracted from *B. oleracea* (pH-sensitive) as an alternative in monitoring ocular physiology. The colorimetric response was contrasted with delphinidin chloride (commercial dye) to obtain the proper concentration to functionalize contact lenses. Huang et al. [80] developed a novel type of conductive and biocompatible neuronal interface based on a nanogel composed of chitosan-modified amphiphilic poly(3,4-ethylenedioxythiophene), which has biomimetic properties. They proposed an alternative to deep brain electrical stimulation combined with manganese-enhanced magnetic resonance imaging.

Some monitoring devices require polymeric coating using similar materials to the mentioned for point-of-care devices (see Table 3.), such as PEDOT coatings [15].

4.3   Treatment and therapy

4.3.1 Drug delivery

The controlled release of pharmaceutical compounds has been improved by the use of polymeric materials, finding more effective administration methods, for example Li et al. [81] obtained prepared a chitosan film that had rapid release and high permeability of the cornea. Their chitosan was loaded with brimonidine tartrate (as a model drug), was stable and had good mucoadhesion and biocompatibility properties. Other example is the use of Poly[(d, l)-lactide-co-glycolide] (PLGA) for brain drug delivery; in the form of nanoparticles for neural interfaces and heat-sensitive drug loads [82] or as nanofiber films loaded with lidocaine and the protein Epidermal growth factor [83]. Those implants and other polymeric materials respond to stimuli of the change of a specific property.

4.3.2 Surgical applications

Plymale et al. [84] conducted a prospective pilot analysis to study quality of life of patients undergoing ventral and incisional hernia repair (VIHR) with P4HB (protein) mesh and defined perioperative characteristics with a follow-up of 24 months after the operation. Surgical site incidents occurred in 19% of patients, most of which were seroma.

Considering a baseline of components physical and emotional, quality of life of the patients was significantly improved after 24 months.

4.3.3 Implants and prosthetics

A common prosthetic application is the use of contact or intraocular lenses. Although lenses are generally used to correct vision problems, recent works [85,86,87] aim to provide additional features that aid comfort. As is the case of Mehta et al. [85], who developed a one-step sonochemical coating using nanoparticles of ZnO, chitosan and gallic acid for contact lenses. Their aim was providing antioxidant properties, improved wettability, and antimicrobial properties while maintaining lens geometric and refractive properties. Some devices are destined to fulfil aesthetic functions, such as composites based on porous silicon nitride ceramics with infiltration of poly (methyl methacrylate) for dental applications [88], and PLA/Al_2O_3 nano-scaffolds or chitosan as dental resins [89,90,91].

Other devices are essential for sustaining life such the manufacturing of neural implants based on photosensitive polyimide (PSPI) and PEG [92], aortic valves with polymeric tissue [93,94], heart valve replacements with poly(urethane-urea siloxane) [95], blood vessels [96] and bone implants for cranial reconstruction with polymethylmethacrylate (PMMA) [97]. Cranial reconstruction can be performed using computer-aided design and manufacturing techniques to produce models that are implanted easily without more invasive techniques. 3D printing technology has also made it possible to develop prostheses with very well defined structures, particularly based on polymers, due to the ease of printing [98].

4.3.4 Tissue engineering and healing

As tissue engineering aims to replace biological functions, the materials involved must comply with biocompatibility, degradability, cytotoxicity, immunotoxicity, moisture retention, and promotion of cell proliferation activity requirements. The advances are development of scaffold based on nanofibers using polylactic acid, thermoplastic polyurethane, poly(vinyl pyrrolidone)-iodine or polycaprolactone [99,100], thin films of chitosan-based polymer matrix for tissue culture [101], and antifungal and antibacterial hydrogels for wound dressings using poly (AA-co-AAm), PVA or PHEMA [102,103] [104]. Other materials that can be used for wound dressing or as medical adhesives [105] [106,107,108], as the studied by Zheng et al. [109], that manufactured quaternized catechol-modified chitosan and incorporated it into poly(d,l-lactide)-poly(ethylene glycol)-poly(d,l-lactide), an injectable hydrogel that is sensitive to heat and with antibacterial and tissue adhesive properties.

Materials Research Forum LLC
https://doi.org/10.21741/9781644901892-1

4.3.5 Coatings and membranes

Membranes are structures that play an important role in clinical medicine for dialysis systems. Examples of materials in this area: PLA [110], copolymers (PLA-PHEMA) [111], and poly(aryl ether sulfone)-poly(vinylpyrrolidone) (PAES-PVP) [112].

Coatings are often used for orthopedic applications, for instance a chitosan nanogel for dental plaque that could be used as part of toothpaste formulations [113], or a composite coating of gelatin/chitosan nanospheres [114]. These polymeric composite coatings are able to improve the corrosion resistance of dental devices or instruments.

Conclusions

This chapter provides an insight into the enormous number of applications that are available for using polymers in the clinical area. Polymers physico-chemical properties can be modified by changing synthesis conditions or being chemically modified. Criteria for make polymers suitable for medical purposes can be summarized in the versatility of polymers to be modified varying parameters (i.e. tacticity, molecular weight, functional groups, crystallinity, crosslinking degree, and others) and the compatibility with living beings or biomolecules combined with the ease of synthesis and processability of most of them. For example, polysaccharides are preferred due to their biocompatibility and low toxicity. Recent advances show non-traditional applications for polymers, before considered only for ceramics or metals, as it can be seen in conductive polymers as PEDOT, PPY y PANI, which are employed for electrical cell stimuli or organic sensor development. Thermal behaviour and pH response properties are commonly employed in controlled drug release and local therapy. Some examples of these polymers are PNIPAM and CPMA respectively. Polymers for device development in medicine must have features suitable for each application such as prevention and diagnosis, monitoring, and treatment and therapy. Polymers are also commonly employed in the development of coatings, membranes and textiles with prominent performance. Examples include antibacterial coatings, dialysis membranes and wound dressing applications. Certainly, the use of polymeric materials in the field of clinical medicine covers countless applications, the scope of which is expanding by leaps and bounds as knowledge grows through the birth of new discoveries and inventions from the mind of researchers. These advances in the development and research of novel materials are leading to personalized medicine.

References

[1] D. Hutmacher, M.B. Hürzeler, H. Schliephake, A review of material properties of biodegradable and bioresorbable polymers and devices for GTR and GBR applications., Int. J. Oral Maxillofac. Implants. 11 (1996).

[2] C.L. Burrell, S.M. Canavan, M.M. Emery, J.C. Ohman, Broken bones: Trauma analysis on a medieval population from Poulton, Cheshire, Explor. Mediev. Cult. Ser. (2018) 71–91. https://doi.org/10.1163/9789004363786_005

[3] F.S. Shahar, M.T.H. Sultan, S.H. Lee, M. Jawaid, A.U.M. Shah, S.N.A. Safri, P.N. Sivasankaran, A review on the orthotics and prosthetics and the potential of kenaf composites as alternative materials for ankle-foot orthosis, J. Mech. Behav. Biomed. Mater. 99 (2019) 169–185. https://doi.org/10.1016/j.jmbbm.2019.07.020

[4] Y.-J. Chen, H. Lin, X. Zhang, W. Huang, L. Shi, D. Wang, Application of 3D–printed and patient-specific cast for the treatment of distal radius fractures: initial experience, 3D Print. Med. 3 (2017) 11. https://doi.org/10.1186/s41205-017-0019-y

[5] P.S.P. Poh, M.A. Woodruff, E. Garc\'ıa-Gareta, Polymer-based composites for musculoskeletal regenerative medicine, in: Biomater. Organ Tissue Regen., Elsevier, 2020: pp. 33–82. https://doi.org/10.1016/B978-0-08-102906-0.00003-9

[6] M.F. Maitz, Applications of synthetic polymers in clinical medicine, Biosurface Biotribology. 1 (2015) 161–176. https://doi.org/10.1016/j.bsbt.2015.08.002

[7] A.P. Vieira, A.F.R. Pimenta, D. Silva, M.H. Gil, P. Alves, P. Coimbra, J.L.G.C. Mata, D. Bozukova, T.R. Correia, I.J. Correia, Others, Surface modification of an intraocular lens material by plasma-assisted grafting with 2-hydroxyethyl methacrylate (HEMA), for controlled release of moxifloxacin, Eur. J. Pharm. Biopharm. 120 (2017) 52–62. https://doi.org/10.1016/j.ejpb.2017.08.006

[8] G.B. Jung, K.-H. Jin, H.-K. Park, Physicochemical and surface properties of acrylic intraocular lenses and their clinical significance, J. Pharm. Investig. 47 (2017) 453–460. https://doi.org/10.1007/s40005-017-0323-y

[9] D. Lee, S. Cho, H.S. Park, I. Kwon, Ocular drug delivery through pHEMA-Hydrogel contact lenses Co-loaded with lipophilic vitamins, Sci. Rep. 6 (2016) 1–8. https://doi.org/10.1038/s41598-016-0001-8

[10] C. Sun, Y.-X. Wang, M. Sun, Y. Zou, C. Zhang, S. Cheng, W. Hu, Facile and cost-effective liver cancer diagnosis by water-gated organic field-effect transistors, Biosens. Bioelectron. 164 (2020) 112251. https://doi.org/10.1016/j.bios.2020.112251

[11]   N. Wang, A. Yang, Y. Fu, Y. Li, F. Yan, Functionalized organic thin film transistors for biosensing, Acc. Chem. Res. 52 (2019) 277–287. https://doi.org/10.1021/acs.accounts.8b00448

[12]   K. Liang, Y. Gao, S. Xiao, F.R. Tay, M.D. Weir, X. Zhou, T.W. Oates, C. Zhou, J. Li, H.H.K. Xu, Poly (amido amine) and rechargeable adhesive containing calcium phosphate nanoparticles for long-term dentin remineralization, J. Dent. 85 (2019) 47–56. https://doi.org/10.1016/j.jdent.2019.04.011

[13]   E.E. Totu, A.C. Nechifor, G. Nechifor, H.Y. Aboul-Enein, C.M. Cristache, Poly (methyl methacrylate) with TiO2 nanoparticles inclusion for stereolitographic complete denture manufacturing- the fututre in dental care for elderly edentulous patients?, J. Dent. 59 (2017) 68–77. https://doi.org/10.1016/j.jdent.2017.02.012

[14]   T. Lenz-Habijan, P. Bhogal, M. Peters, A. Bufe, R.M. Moreno, C. Bannewitz, H. Monstadt, H. Henkes, Hydrophilic stent coating inhibits platelet adhesion on stent surfaces: initial results in vitro, Cardiovasc. Intervent. Radiol. 41 (2018) 1779–1785. https://doi.org/10.1007/s00270-018-2036-7

[15]   C. Bodart, N. Rossetti, J. Hagler, P. Chevreau, D. Chhin, F. Soavi, S.B. Schougaard, F. Amzica, F. Cicoira, Electropolymerized poly (3, 4-ethylenedioxythiophene)(PEDOT) coatings for implantable deep-brain-stimulating microelectrodes, ACS Appl. Mater. Interfaces. 11 (2019) 17226–17233. https://doi.org/10.1021/acsami.9b03088

[16]   A. Golabchi, B. Wu, B. Cao, C.J. Bettinger, X.T. Cui, Zwitterionic polymer/polydopamine coating reduce acute inflammatory tissue responses to neural implants, Biomaterials. 225 (2019) 119519. https://doi.org/10.1016/j.biomaterials.2019.119519

[17]   R.L. Li, J. Russ, C. Paschalides, G. Ferrari, H. Waisman, J.W. Kysar, D. Kalfa, Mechanical considerations for polymeric heart valve development: Biomechanics, materials, design and manufacturing, Biomaterials. 225 (2019) 119493. https://doi.org/10.1016/j.biomaterials.2019.119493

[18]   J.R. Stasiak, M. Serrani, E. Biral, J.V. Taylor, A.G. Zaman, S. Jones, T. Ness, F. De Gaetano, M.L. Costantino, V.D. Bruno, Others, Design, development, testing at ISO standards and in vivo feasibility study of a novel polymeric heart valve prosthesis, Biomater. Sci. 8 (2020) 4467–4480. https://doi.org/10.1039/D0BM00412J

[19]   E. Ovcharenko, M. Rezvova, P. Nikishau, S. Kostjuk, T. Glushkova, L. Antonova, D. Trebushat, T. Akentieva, D. Shishkova, E. Krivikina, Others, Polyisobutylene-

Materials Research Forum LLC
https://doi.org/10.21741/9781644901892-1

based thermoplastic elastomers for manufacturing polymeric heart valve leaflets: In vitro and in vivo results, Appl. Sci. 9 (2019) 4773. https://doi.org/10.3390/app9224773

[20] H. Akai, K. Shiraishi, M. Yokoyama, K. Yasaka, M. Nojima, Y. Inoue, O. Abe, K. Ohtomo, S. Kiryu, PEG-poly (L-lysine)-based polymeric micelle MRI contrast agent: Feasibility study of a Gd-micelle contrast agent for MR lymphography, J. Magn. Reson. Imaging. 47 (2018) 238–245. https://doi.org/10.1002/jmri.25740

[21] P. Lei, R. An, P. Zhang, S. Yao, S. Song, L. Dong, X. Xu, K. Du, J. Feng, H. Zhang, Ultrafast synthesis of ultrasmall poly (Vinylpyrrolidone)-protected Bismuth Nanodots as a multifunctional theranostic agent for in vivo dual-modal CT/Photothermal-imaging-guided photothermal therapy, Adv. Funct. Mater. 27 (2017) 1702018. https://doi.org/10.1002/adfm.201702018

[22] O. Ter Beek, D. Pavlenko, M. Suck, S. Helfrich, L. Bolhuis-Versteeg, D. Snisarenko, C. Causserand, P. Bacchin, P. Aimar, R. van Oerle, Others, New membranes based on polyethersulfone–SlipSkin™ polymer blends with low fouling and high blood compatibility, Sep. Purif. Technol. 225 (2019) 60–73. https://doi.org/10.1016/j.seppur.2019.05.049

[23] A. Boschetti-de-Fierro, W. Beck, H. Hildwein, B. Krause, M. Storr, C. Zweigart, Membrane innovation in dialysis, Expand. Hemodial. 191 (2017) 100–114. https://doi.org/10.1159/000479259

[24] P. Ganesan, Natural and bio polymer curative films for wound dressing medical applications, Wound Med. 18 (2017) 33–40. https://doi.org/10.1016/j.wndm.2017.07.002

[25] E.A. Kamoun, E.-R.S. Kenawy, X. Chen, A review on polymeric hydrogel membranes for wound dressing applications: PVA-based hydrogel dressings, J. Adv. Res. 8 (2017) 217–233. https://doi.org/10.1016/j.jare.2017.01.005

[26] P. Khandelwal, A. Das, C.K. Sen, S.P. Srinivas, S. Roy, S. Khanna, A surfactant polymer wound dressing protects human keratinocytes from inducible necroptosis, Sci. Rep. 11 (2021) 1–15. https://doi.org/10.1038/s41598-021-82260-x

[27] R. Fu, W. Luo, R. Nazempour, D. Tan, H. Ding, K. Zhang, L. Yin, J. Guan, X. Sheng, Implantable and biodegradable poly (L-lactic acid) fibers for optical neural interfaces, Adv. Opt. Mater. 6 (2018) 1700941. https://doi.org/10.1002/adom.201700941

[28] R. Ramachandran, V.R. Junnuthula, G.S. Gowd, A. Ashokan, J. Thomas, R. Peethambaran, A. Thomas, A.K.K. Unni, D. Panikar, S.V. Nair, Others, Theranostic 3-

Dimensional nano brain-implant for prolonged and localized treatment of recurrent glioma, Sci. Rep. 7 (2017) 1–16. https://doi.org/10.1038/s41598-016-0028-x

[29]   Y. Zhou, L. Kang, Z. Yue, X. Liu, G.G. Wallace, Composite tissue adhesive containing catechol-modified hyaluronic acid and poly-l-lysine, ACS Appl. Bio Mater. 3 (2019) 628–638. https://doi.org/10.1021/acsabm.9b01003

[30]   B. Peng, X. Lai, L. Chen, X. Lin, C. Sun, L. Liu, S. Qi, Y. Chen, K.W. Leong, Scarless wound closure by a mussel-inspired poly (amidoamine) tissue adhesive with tunable degradability, ACS Omega. 2 (2017) 6053–6062. https://doi.org/10.1021/acsomega.7b01221

[31]   D. Mendes Junior, J.A. Domingues, M.A. Hausen, S.M.M. Cattani, A. Aragones, A.L.R. Oliveira, R.F. Inácio, M.L.P. Barbo, E.A.R. Duek, Study of mesenchymal stem cells cultured on a poly (lactic-co-glycolic acid) scaffold containing simvastatin for bone healing, J. Appl. Biomater. Funct. Mater. 15 (2017) 133–141. https://doi.org/10.5301/jabfm.5000338

[32]   S. Srisang, N. Wongsuwan, A. Boongird, M. Ungsurungsie, P. Wanasawas, N. Nasongkla, Multilayer nanocoating of Foley urinary catheter by chlorhexidine-loaded nanoparticles for prolonged release and anti-infection of urinary tract, Int. J. Polym. Mater. Polym. Biomater. 69 (2020) 1081–1089. https://doi.org/10.1080/00914037.2019.1655752

[33]   S. Kariya, M. Nakatani, Y. Ono, T. Maruyama, Y. Ueno, A. Komemushi, N. Tanigawa, Assessment of the antithrombogenicity of a poly-2-methoxyethylacrylate-coated central venous port-catheter system, Cardiovasc. Intervent. Radiol. (2020) 1–6. https://doi.org/10.1177/1129729820983175

[34]   P. Ahmadi, N. Nazeri, M.A. Derakhshan, H. Ghanbari, Preparation and characterization of polyurethane/chitosan/CNT nanofibrous scaffold for cardiac tissue engineering, Int. J. Biol. Macromol. 180 (2021) 590–598. https://doi.org/10.1016/j.ijbiomac.2021.03.001

[35]   N. Celikkin, S. Mastrogiacomo, J. Jaroszewicz, X.F. Walboomers, W. Swieszkowski, Gelatin methacrylate scaffold for bone tissue engineering: the influence of polymer concentration, J. Biomed. Mater. Res. A. 106 (2018) 201–209. https://doi.org/10.1002/jbm.a.36226

[36]   Y. Zhang, K. Kang, N. Zhu, G. Li, X. Zhou, A. Zhang, Q. Yi, Y. Wu, Bottlebrush-like highly efficient antibacterial coating constructed using α-helical peptide dendritic polymers on the poly (styrene-b-(ethylene-co-butylene)-b-styrene) surface, J. Mater. Chem. B. 8 (2020) 7428–7437. https://doi.org/10.1039/D0TB01336F

[37]   H. Gulliksson, S. Meinke, A. Ravizza, L. Larsson, P. Höglund, Storage of red blood cells in a novel polyolefin blood container: a pilot in vitro study, Vox Sang. 112 (2017) 33–39. https://doi.org/10.1111/vox.12472

[38]   S. Sarigul Guduk, N. Karaca, Safety and complications of absorbable threads made of poly-L-lactic acid and poly lactide/glycolide: experience with 148 consecutive patients, J. Cosmet. Dermatol. 17 (2018) 1189–1193. https://doi.org/10.1111/jocd.12519

[39]   F. López-Saucedo, G.G. Flores-Rojas, E. Bucio, C. Alvarez-Lorenzo, A. Concheiro, O. González-Antonio, Achieving antimicrobial activity through poly (N-methylvinylimidazolium) iodide brushes on binary-grafted polypropylene suture threads, MRS Commun. 7 (2017) 938–946. https://doi.org/10.1557/mrc.2017.121

[40]   S.J. Lee, H.H. Jo, K.S. Lim, D. Lim, S. Lee, J.H. Lee, W.D. Kim, M.H. Jeong, J.Y. Lim, I.K. Kwon, Others, Heparin coating on 3D printed poly (l-lactic acid) biodegradable cardiovascular stent via mild surface modification approach for coronary artery implantation, Chem. Eng. J. 378 (2019) 122116. https://doi.org/10.1016/j.cej.2019.122116

[41]   K. Wang, Q. Zhang, L. Zhao, Y. Pan, T. Wang, D. Zhi, S. Ma, P. Zhang, T. Zhao, S. Zhang, Others, Functional modification of electrospun poly (ε-caprolactone) vascular grafts with the fusion protein VEGF–HGFI enhanced vascular regeneration, ACS Appl. Mater. Interfaces. 9 (2017) 11415–11427. https://doi.org/10.1021/acsami.6b16713

[42]   M. Rahman, M.R. Hasan, Synthetic biopolymers, Funct. Biopolym. Ed. Jafar Mazumder MA Shear. H Al-Ahmed Ed. Springer Int. Publ. (2019) 1–43.

[43]   B. Nagel, H. Dellweg, L.M. Gierasch, Glossary for chemists of terms used in biotechnology (IUPAC Recommendations 1992), Pure Appl. Chem. 64 (1992) 143–168. https://doi.org/10.1351/pac199264010143

[44]   Y. Nishio, Material functionalization of cellulose and related polysaccharides via diverse microcompositions, Polysacch. II. (2006) 97–151. https://doi.org/10.1007/12_095

[45]   E. Pinho, G. Soares, Functionalization of cotton cellulose for improved wound healing, J. Mater. Chem. B. 6 (2018) 1887–1898. https://doi.org/10.1039/C8TB00052B

[46]   H. Mertaniemi, C. Escobedo-Lucea, A. Sanz-Garcia, C. Gandía, A. Mäkitie, J. Partanen, O. Ikkala, M. Yliperttula, Human stem cell decorated nanocellulose threads

Materials Research Forum LLC
https://doi.org/10.21741/9781644901892-1

for biomedical applications, Biomaterials. 82 (2016) 208–220.
https://doi.org/10.1016/j.biomaterials.2015.12.020

[47]   R.O. Ebewele, Polymerization Mechanisms, in: Polym. Sci. Technol., CRC Press, 2000.

[48]   M.C. Hacker, J. Krieghoff, A.G. Mikos, Chapter 33 - Synthetic Polymers, in: A. Atala, R. Lanza, A.G. Mikos, R. Nerem (Eds.), Princ. Regen. Med. Third Ed., Academic Press, Boston, 2019: pp. 559–590. https://doi.org/10.1016/B978-0-12-809880-6.00033-3

[49]   A. Kulcke, C. Gurschler, G. Spöck, R. Leitner, M. Kraft, On-Line Classification of Synthetic Polymers Using near Infrared Spectral Imaging, J. Infrared Spectrosc. 11 (2003) 71–81. https://doi.org/10.1255/jnirs.355

[50]   M. Movahedi, A. Asefnejad, M. Rafienia, M.T. Khorasani, Potential of novel electrospun core-shell structured polyurethane/starch (hyaluronic acid) nanofibers for skin tissue engineering: In vitro and in vivo evaluation, Int. J. Biol. Macromol. 146 (2020) 627–637. https://doi.org/10.1016/j.ijbiomac.2019.11.233

[51]   S.M. Bittner, B.T. Smith, L. Diaz-Gomez, C.D. Hudgins, A.J. Melchiorri, D.W. Scott, J.P. Fisher, A.G. Mikos, Fabrication and mechanical characterization of 3D printed vertical uniform and gradient scaffolds for bone and osteochondral tissue engineering, Acta Biomater. 90 (2019) 37–48. https://doi.org/10.1016/j.actbio.2019.03.041

[52]   S. Mondal, T.P. Nguyen, G. Hoang, P. Manivasagan, M.H. Kim, S.Y. Nam, J. Oh, Others, Hydroxyapatite nano bioceramics optimized 3D printed poly lactic acid scaffold for bone tissue engineering application, Ceram. Int. 46 (2020) 3443–3455. https://doi.org/10.1016/j.ceramint.2019.10.057

[53]   L. Vogt, L.R. Rivera, L. Liverani, A. Piegat, M. El Fray, A.R. Boccaccini, Poly (ε-caprolactone)/poly (glycerol sebacate) electrospun scaffolds for cardiac tissue engineering using benign solvents, Mater. Sci. Eng. C. 103 (2019) 109712. https://doi.org/10.1016/j.msec.2019.04.091

[54]   H. Li, F. Cheng, C. Chávez-Madero, J. Choi, X. Wei, X. Yi, T. Zheng, J. He, Manufacturing and physical characterization of absorbable oxidized regenerated cellulose braided surgical sutures, Int. J. Biol. Macromol. 134 (2019) 56–62. https://doi.org/10.1016/j.ijbiomac.2019.05.030

[55]   R. Hua, Y. Tian, J. Cheng, G. Wu, W. Jiang, Z. Ni, G. Zhao, The effect of intrinsic characteristics on mechanical properties of poly (l-lactic acid) bioresorbable vascular

stents, Med. Eng. Phys. 81 (2020) 118–124.
https://doi.org/10.1016/j.medengphy.2020.04.006

[56] P.R. Neumann, D.L. Crossley, M. Turner, M. Ingleson, M. Green, J. Rao, L.A.
Dailey, In vivo optical performance of a new class of near-infrared-emitting
conjugated polymers: borylated PF8-BT, ACS Appl. Mater. Interfaces. 11 (2019)
46525–46535. https://doi.org/10.1021/acsami.9b17022

[57] D.N. Heo, S.-J. Lee, R. Timsina, X. Qiu, N.J. Castro, L.G. Zhang, Development of
3D printable conductive hydrogel with crystallized PEDOT: PSS for neural tissue
engineering, Mater. Sci. Eng. C. 99 (2019) 582–590.
https://doi.org/10.1016/j.msec.2019.02.008

[58] S. Kumar, M. Umar, A. Saifi, S. Kumar, S. Augustine, S. Srivastava, B.D.
Malhotra, Electrochemical paper based cancer biosensor using iron oxide
nanoparticles decorated PEDOT: PSS, Anal. Chim. Acta. 1056 (2019) 135–145.
https://doi.org/10.1016/j.aca.2018.12.053

[59] M. Aydin, E.B. Aydin, M.K. Sezgintürk, A highly selective electrochemical
immunosensor based on conductive carbon black and star PGMA polymer composite
material for IL-8 biomarker detection in human serum and saliva, Biosens.
Bioelectron. 117 (2018) 720–728. https://doi.org/10.1016/j.bios.2018.07.010

[60] M. Tertiş, B. Ciui, M. Suciu, R. S\uandulescu, C. Cristea, Label-free
electrochemical aptasensor based on gold and polypyrrole nanoparticles for interleukin
6 detection, Electrochimica Acta. 258 (2017) 1208–1218.
https://doi.org/10.1016/j.electacta.2017.11.176

[61] M. Müller, B. Urban, B. Reis, X. Yu, A.L. Grab, E.A. Cavalcanti-Adam, D.
Kuckling, Switchable Release of Bone Morphogenetic Protein from Thermoresponsive
Poly (NIPAM-co-DMAEMA)/Cellulose Sulfate Particle Coatings, Polymers. 10
(2018) 1314. https://doi.org/10.3390/polym10121314

[62] K. Brewer, B. Gundsambuu, P. Facal Marina, S.C. Barry, A. Blencowe,
Thermoresponsive poly (ε-Caprolactone)-poly (Ethylene/Propylene Glycol)
copolymers as injectable hydrogels for cell therapies, Polymers. 12 (2020) 367.
https://doi.org/10.3390/polym12020367

[63] T. Xu, Y. Ma, J. Huang, H. Lai, D. Yuan, X. Tang, L. Yang, Self-organized
thermo-responsive poly (lactic-co-glycolic acid)-graft-pullulan nanoparticles for
synergistic thermo-chemotherapy of tumor, Carbohydr. Polym. 237 (2020) 116104.
https://doi.org/10.1016/j.carbpol.2020.116104

[64] Y.-T. Qin, H. Peng, X.-W. He, W.-Y. Li, Y.-K. Zhang, pH-Responsive polymer-stabilized ZIF-8 nanocomposites for fluorescence and magnetic resonance dual-modal imaging-guided chemo-/photodynamic combinational cancer therapy, ACS Appl. Mater. Interfaces. 11 (2019) 34268–34281. https://doi.org/10.1021/acsami.9b12641

[65] A. Roointan, J. Farzanfar, S. Mohammadi-Samani, A. Behzad-Behbahani, F. Farjadian, Smart pH responsive drug delivery system based on poly (HEMA-co-DMAEMA) nanohydrogel, Int. J. Pharm. 552 (2018) 301–311. https://doi.org/10.1016/j.ijpharm.2018.10.001

[66] Z. Qin, T. Chen, W. Teng, Q. Jin, J. Ji, Mixed-charged zwitterionic polymeric micelles for tumor acidic environment responsive intracellular drug delivery, Langmuir. 35 (2018) 1242–1248. https://doi.org/10.1021/acs.langmuir.8b00471

[67] E. Bihar, S. Wustoni, A.M. Pappa, K.N. Salama, D. Baran, S. Inal, A fully inkjet-printed disposable glucose sensor on paper, Npj Flex. Electron. 2 (2018) 1–8. https://doi.org/10.1038/s41528-018-0044-y

[68] B.K. Shrestha, R. Ahmad, S. Shrestha, C.H. Park, C.S. Kim, Globular shaped polypyrrole doped well-dispersed functionalized multiwall carbon nanotubes/nafion composite for enzymatic glucose biosensor application, Sci. Rep. 7 (2017) 1–13. https://doi.org/10.1038/s41598-016-0028-x

[69] A. Khodadadi, E. Faghih-Mirzaei, H. Karimi-Maleh, A. Abbaspourrad, S. Agarwal, V.K. Gupta, A new epirubicin biosensor based on amplifying DNA interactions with polypyrrole and nitrogen-doped reduced graphene: experimental and docking theoretical investigations, Sens. Actuators B Chem. 284 (2019) 568–574. https://doi.org/10.1016/j.snb.2018.12.164

[70] J. Wang, N. Hui, Zwitterionic poly (carboxybetaine) functionalized conducting polymer polyaniline nanowires for the electrochemical detection of carcinoembryonic antigen in undiluted blood serum, Bioelectrochemistry. 125 (2019) 90–96. https://doi.org/10.1016/j.bioelechem.2018.09.006

[71] S.M. Sheta, S.M. El-Sheikh, D.I. Osman, A.M. Salem, O.I. Ali, F.A. Harraz, W.G. Shousha, M.A. Shoeib, S.M. Shawky, D.D. Dionysiou, A novel HCV electrochemical biosensor based on a polyaniline@ Ni-MOF nanocomposite, Dalton Trans. 49 (2020) 8918–8926. https://doi.org/10.1039/D0DT01408G

[72] A.R. Rebelo, C. Liu, K.-H. Schäfer, M. Saumer, G. Yang, Y. Liu, Poly (4-vinylaniline)/polyaniline bilayer-functionalized bacterial cellulose for flexible electrochemical biosensors, Langmuir. 35 (2019) 10354–10366. https://doi.org/10.1021/acs.langmuir.9b01425

[73] M. Xu, Y. Song, Y. Ye, C. Gong, Y. Shen, L. Wang, L. Wang, A novel flexible electrochemical glucose sensor based on gold nanoparticles/polyaniline arrays/carbon cloth electrode, Sens. Actuators B Chem. 252 (2017) 1187–1193. https://doi.org/10.1016/j.snb.2017.07.147

[74] X. Huang, W. Shi, J. Li, N. Bao, C. Yu, H. Gu, Determination of salivary uric acid by using poly (3, 4-ethylenedioxythipohene) and graphene oxide in a disposable paper-based analytical device, Anal. Chim. Acta. 1103 (2020) 75–83. https://doi.org/10.1016/j.aca.2019.12.057

[75] K. Qu, S.M. Kondengaden, J. Li, X. Wang, M.D. Sevilla, L. Li, X. Zeng, Carbohydrate-functionalized polythiophene biointerface: design, fabrication, characterization and application for protein analysis, Appl. Surf. Sci. 486 (2019) 561–570. https://doi.org/10.1016/j.apsusc.2019.04.231

[76] M. Kim, R. Iezzi Jr, B.S. Shim, D.C. Martin, Impedimetric biosensors for detecting vascular endothelial growth factor (VEGF) based on poly (3, 4-ethylene dioxythiophene)(PEDOT)/gold nanoparticle (Au NP) composites, Front. Chem. 7 (2019) 234. https://doi.org/10.3389/fchem.2019.00234

[77] L. Ma, S. Jayachandran, Z. Li, Z. Song, W. Wang, X. Luo, Antifouling and conducting PEDOT derivative grafted with polyglycerol for highly sensitive electrochemical protein detection in complex biological media, J. Electroanal. Chem. 840 (2019) 272–278. https://doi.org/10.1016/j.jelechem.2019.04.002

[78] S. Faalnouri, D. Çimen, N. Bereli, A. Denizli, Surface Plasmon Resonance Nanosensors for Detecting Amoxicillin in Milk Samples with Amoxicillin Imprinted Poly (hydroxyethyl methacrylate-N-methacryloyl-(L)-glutamic acid), ChemistrySelect. 5 (2020) 4761–4769. https://doi.org/10.1002/slct.202000621

[79] R.S. Riaz, M. Elsherif, R. Moreddu, I. Rashid, M.U. Hassan, A.K. Yetisen, H. Butt, Anthocyanin-functionalized contact lens sensors for ocular pH monitoring, ACS Omega. 4 (2019) 21792–21798. https://doi.org/10.1021/acsomega.9b02638

[80] W.-C. Huang, Y.-C. Lo, C.-Y. Chu, H.-Y. Lai, Y.-Y. Chen, S.-Y. Chen, Conductive nanogel-interfaced neural microelectrode arrays with electrically controlled in-situ delivery of manganese ions enabling high-resolution MEMRI for synchronous neural tracing with deep brain stimulation, Biomaterials. 122 (2017) 141–153. https://doi.org/10.1016/j.biomaterials.2017.01.013

[81] B. Li, J. Wang, Q. Gui, H. Yang, Drug-loaded chitosan film prepared via facile solution casting and air-drying of plain water-based chitosan solution for ocular drug

delivery, Bioact. Mater. 5 (2020) 577–583.
https://doi.org/10.1016/j.bioactmat.2020.04.013

[82]   A.D. Holmkvist, J. Agorelius, M. Forni, U.J. Nilsson, C.E. Linsmeier, J. Schouenborg, Local delivery of minocycline-loaded PLGA nanoparticles from gelatin-coated neural implants attenuates acute brain tissue responses in mice, J. Nanobiotechnology. 18 (2020) 1–12. https://doi.org/10.1186/s12951-020-0585-9

[83]   C.-W. Kao, Y.-Y. Tseng, K.-S. Liu, Y.-W. Liu, J.-C. Chen, H.-L. He, Y.-C. Kau, S.-J. Liu, Anesthetics and human epidermal growth factor incorporated into anti-adhesive nanofibers provide sustained pain relief and promote healing of surgical wounds, Int. J. Nanomedicine. 14 (2019) 4007. https://doi.org/10.2147/IJN.S202402

[84]   M.A. Plymale, D.L. Davenport, A. Dugan, A. Zachem, J.S. Roth, Ventral hernia repair with poly-4-hydroxybutyrate mesh, Surg. Endosc. 32 (2018) 1689–1694. https://doi.org/10.1007/s00464-017-5848-7

[85]   P. Mehta, A.A. Al-Kinani, M.S. Arshad, N. Singh, S.M. van der Merwe, M.-W. Chang, R.G. Alany, Z. Ahmad, Engineering and development of chitosan-based Nanocoatings for Ocular Contact Lenses, J. Pharm. Sci. 108 (2019) 1540–1551. https://doi.org/10.1016/j.xphs.2018.11.036

[86]   J. Hoyo, K. Ivanova, E. Guaus, T. Tzanov, Multifunctional ZnO NPs-chitosan-gallic acid hybrid nanocoating to overcome contact lenses associated conditions and discomfort, J. Colloid Interface Sci. 543 (2019) 114–121. https://doi.org/10.1016/j.jcis.2019.02.043

[87]   N.-P.-D. Tran, M.-C. Yang, Others, Synthesis and characterization of silicone contact lenses based on TRIS-DMA-NVP-HEMA hydrogels, Polymers. 11 (2019) 944. https://doi.org/10.3390/polym11060944

[88]   F. Wang, J. Guo, K. Li, J. Sun, Y. Zeng, C. Ning, High strength polymer/silicon nitride composites for dental restorations, Dent. Mater. 35 (2019) 1254–1263. https://doi.org/10.1016/j.dental.2019.05.022

[89]   M. Ranjbar, G. Dehghan Noudeh, M.-A. Hashemipour, I. Mohamadzadeh, A systematic study and effect of PLA/Al2O3 nanoscaffolds as dental resins: mechanochemical properties, Artif. Cells Nanomedicine Biotechnol. 47 (2019) 201–209. https://doi.org/10.1080/21691401.2018.1548472

[90]   M. Ducret, A. Montembault, J. Josse, M. Pasdeloup, A. Celle, R. Benchrih, F. Mallein-Gerin, B. Alliot-Licht, L. David, J.-C. Farges, Design and characterization of

Materials Research Forum LLC
https://doi.org/10.21741/9781644901892-1

a chitosan-enriched fibrin hydrogel for human dental pulp regeneration, Dent. Mater. 35 (2019) 523–533. https://doi.org/10.1016/j.dental.2019.01.018

[91]   A. Xing, Q. Sun, Y. Meng, Y. Zhang, X. Li, B. Han, A hydroxyl-containing hyperbranched polymer as a multi-purpose modifier for a dental epoxy, React. Funct. Polym. 149 (2020) 104505. https://doi.org/10.1016/j.reactfunctpolym.2020.104505

[92]   W. Zhang, X. Zhou, Y. He, L. Xu, J. Xie, Implanting mechanics of PEG/DEX coated flexible neural probe: impacts of fabricating methods, Biomed. Microdevices. 23 (2021) 1–16. https://doi.org/10.1007/s10544-020-00537-w

[93]   O.M. Rotman, B. Kovarovic, W.-C. Chiu, M. Bianchi, G. Marom, M.J. Slepian, D. Bluestein, Novel polymeric valve for transcatheter aortic valve replacement applications: in vitro hemodynamic study, Ann. Biomed. Eng. 47 (2019) 113–125. https://doi.org/10.1007/s10439-018-02119-7

[94]   G. Zhu, M.B. Ismail, M. Nakao, Q. Yuan, J.H. Yeo, Numerical and in-vitro experimental assessment of the performance of a novel designed expanded-polytetrafluoroethylene stentless bi-leaflet valve for aortic valve replacement, Plos One. 14 (2019) e0210780. https://doi.org/10.1371/journal.pone.0210780

[95]   C. Jenney, P. Millson, D.W. Grainger, R. Grubbs, P. Gunatillake, S.J. McCarthy, J. Runt, J. Beith, Assessment of a Siloxane Poly (urethane-urea) Elastomer Designed for Implantable Heart Valve Leaflets, Adv. NanoBiomed Res. 1 (2021) 2000032. https://doi.org/10.1002/anbr.202000032

[96]   N. Tran, A. Le, M. Ho, N. Dang, H.H. Thi Thanh, L. Truong, D.P. Huynh, N.T. Hiep, Polyurethane/polycaprolactone membrane grafted with conjugated linoleic acid for artificial vascular graft application, Sci. Technol. Adv. Mater. 21 (2020) 56–66. https://doi.org/10.1080/14686996.2020.1718549

[97]   C. Unterhofer, C. Wipplinger, M. Verius, W. Recheis, C. Thomé, M. Ortler, Reconstruction of large cranial defects with poly-methyl-methacrylate (PMMA) using a rapid prototyping model and a new technique for intraoperative implant modeling, Neurol. Neurochir. Pol. 51 (2017) 214–220. https://doi.org/10.1016/j.pjnns.2017.02.007

[98]   Y. Jang, J.-Y. Sim, J.-K. Park, W.-C. Kim, H.-Y. Kim, J.-H. Kim, Accuracy of 3-unit fixed dental prostheses fabricated on 3D-printed casts, J. Prosthet. Dent. 123 (2020) 135–142. https://doi.org/10.1016/j.prosdent.2018.11.004

[99]   E. Fallahiarezoudar, M. Ahmadipourroudposht, A. Idris, N.M. Yusof, Optimization and development of Maghemite (γ-Fe2O3) filled poly-L-lactic acid

(PLLA)/thermoplastic polyurethane (TPU) electrospun nanofibers using Taguchi orthogonal array for tissue engineering heart valve, Mater. Sci. Eng. C. 76 (2017) 616–627. https://doi.org/10.1016/j.msec.2017.03.120

[100] A.A. Shitole, P. Raut, P. Giram, P. Rade, A. Khandwekar, B. Garnaik, N. Sharma, Poly (vinylpyrrolidone)-iodine engineered poly (ε-caprolactone) nanofibers as potential wound dressing materials, Mater. Sci. Eng. C. 110 (2020) 110731. https://doi.org/10.1016/j.msec.2020.110731

[101] B. Kaczmarek, O. Mazur, O. Mi\lek, M. Michalska-Sionkowska, A.M. Osyczka, K. Kleszczyński, Development of tannic acid-enriched materials modified by poly (ethylene glycol) for potential applications as wound dressing, Prog. Biomater. 9 (2020) 115–123. https://doi.org/10.1007/s40204-020-00136-1

[102] A. Olad, M. Eslamzadeh, F. Katiraee, A. Mirmohseni, Evaluation of in vitro anti-fungal properties of allicin loaded ion cross-linked poly (AA-co-AAm)/PVA/Cloisite 15A Nanocomposite hydrogel films as wound dressing materials, J. Polym. Res. 27 (2020) 1–10. https://doi.org/10.1007/s10965-020-02072-x

[103] G. Tao, Y. Wang, R. Cai, H. Chang, K. Song, H. Zuo, P. Zhao, Q. Xia, H. He, Design and performance of sericin/poly (vinyl alcohol) hydrogel as a drug delivery carrier for potential wound dressing application, Mater. Sci. Eng. C. 101 (2019) 341–351. https://doi.org/10.1016/j.msec.2019.03.111

[104] Z. Di, Z. Shi, M.W. Ullah, S. Li, G. Yang, A transparent wound dressing based on bacterial cellulose whisker and poly (2-hydroxyethyl methacrylate), Int. J. Biol. Macromol. 105 (2017) 638–644. https://doi.org/10.1016/j.ijbiomac.2017.07.075

[105] S. Zhang, H. Li, M. Yuan, M. Yuan, H. Chen, Poly (lactic acid) blends with poly (trimethylene carbonate) as biodegradable medical adhesive material, Int. J. Mol. Sci. 18 (2017) 2041. https://doi.org/10.3390/ijms18102041

[106] W. Chen, R. Wang, T. Xu, X. Ma, Z. Yao, B.O. Chi, H. Xu, A mussel-inspired poly (γ-glutamic acid) tissue adhesive with high wet strength for wound closure, J. Mater. Chem. B. 5 (2017) 5668–5678. https://doi.org/10.1039/C7TB00813A

[107] F. Sun, Y. Bu, Y. Chen, F. Yang, J. Yu, D. Wu, An injectable and instant self-healing medical adhesive for wound sealing, ACS Appl. Mater. Interfaces. 12 (2020) 9132–9140. https://doi.org/10.1021/acsami.0c01022

[108] A. Assmann, A. Vegh, M. Ghasemi-Rad, S. Bagherifard, G. Cheng, E.S. Sani, G.U. Ruiz-Esparza, I. Noshadi, A.D. Lassaletta, S. Gangadharan, Others, A highly

Materials Research Forum LLC
https://doi.org/10.21741/9781644901892-1

adhesive and naturally derived sealant, Biomaterials. 140 (2017) 115–127. https://doi.org/10.1016/j.biomaterials.2017.06.004

[109] Z. Zheng, S. Bian, Z. Li, Z. Zhang, Y. Liu, X. Zhai, H. Pan, X. Zhao, Catechol modified quaternized chitosan enhanced wet adhesive and antibacterial properties of injectable thermo-sensitive hydrogel for wound healing, Carbohydr. Polym. 249 (2020) 116826. https://doi.org/10.1016/j.carbpol.2020.116826

[110] A. Gao, F. Liu, L. Xue, Preparation and evaluation of heparin-immobilized poly (lactic acid)(PLA) membrane for hemodialysis, J. Membr. Sci. 452 (2014) 390–399. https://doi.org/10.1016/j.memsci.2013.10.016

[111] L. Zhu, F. Liu, X. Yu, L. Xue, Poly (lactic acid) hemodialysis membranes with poly (lactic acid)-block-poly (2-hydroxyethyl methacrylate) copolymer as additive: preparation, characterization, and performance, ACS Appl. Mater. Interfaces. 7 (2015) 17748–17755. https://doi.org/10.1021/acsami.5b03951

[112] A. Mollahosseini, S. Argumeedi, A. Abdelrasoul, A. Shoker, A case study of poly (aryl ether sulfone) hemodialysis membrane interactions with human blood: Molecular dynamics simulation and experimental analyses, Comput. Methods Programs Biomed. 197 (2020) 105742. https://doi.org/10.1016/j.cmpb.2020.105742

[113] B. Ashrafi, M. Rashidipour, A. Marzban, S. Soroush, M. Azadpour, S. Delfani, P. Ramak, Mentha piperita essential oils loaded in a chitosan nanogel with inhibitory effect on biofilm formation against S. mutans on the dental surface, Carbohydr. Polym. 212 (2019) 142–149. https://doi.org/10.1016/j.carbpol.2019.02.018

[114] H. Qi, S. Heise, J. Zhou, K. Schuhladen, Y. Yang, N. Cui, R. Dong, S. Virtanen, Q. Chen, A.R. Boccaccini, Others, Electrophoretic deposition of bioadaptive drug delivery coatings on magnesium alloy for bone repair, ACS Appl. Mater. Interfaces. 11 (2019) 8625–8634. https://doi.org/10.1021/acsami.9b01227

Materials Research Forum LLC
https://doi.org/10.21741/9781644901892-2

Chapter 2

The Role of Advanced Polymers in Surgery & Medical Devices

Anupama Rajput[1*], Anamika Debnath[2]

[1]Department of Chemistry, Faculty of Engineering & Technology, Manav Rachna International Institute of Research & Studies, Faridabad, Haryana, India

[2]Department of Chemistry, Daulat Ram College, Delhi University, DELHI India

* anupamarajputchem@gmail.com

Abstract

Polymers find a variety of applications in our day-to-day life. A wide range of synthetic polymers which have become a part of our daily life. Synthetic resin has a variety of applications in our life. Synthetic polymers are uses in craft and biomedical and surgical operation. Medical devices include simple devices; test equipment and implants. Polymers of different types are used extensively in these devices, for weight, cost, and performance benefits. The chapter focuses on some advanced polymers like resin, copolymers and thermoplastics which are therefore used in different fields like medical or surgical in large scale on the basis of structure activity relationship. Advanced polymeric materials and biodegradable polymers and their application in surgical devices were also discussed.

Keywords

Medical Devices, Advanced Polymers, Thermo Plastics, Pacemakers

Contents

1.   Introduction

Polymer science is being progressed exponentially because of growing demands in this advanced world for the production of versatile, sustainable and novel materials for advance applications. In the last few decades, polymer modification techniques have grown in large scale. To meet various purposes such as evaluation, augmentation, replacement of any part of the body, biomaterials like synthetic and natural polymers are popularly used in medical fields [1]. From the second half of the 1960s, the considerable application of synthetic biodegradable polymers introduces a new era in the field of revision surgeries. There has been a constant research going on to make a shift from bio stable materials to biodegradable materials [2-5]. The enormous growth of polymeric biomaterial is in the global talk, for its promising potential in the biomedical applications [6]. The elaborated study over such applications is being mentioned in this chapter.

Different medical applications require different polymeric systems with unique properties. Thus, for this purpose, multifunctional and combinatorial biodegradable polymers have been of great interest.

Material's biocompatibility is greatly influenced by:

    ✓ Its material chemistry

    ✓ Solubility and molecular weight

    ✓ surface energy and lubricity

    ✓ behavior towards water

    ✓ degradation

    ✓ implant's structure and shape [7]

Importantly, excellent biocompatibility of polymers can change significantly over time due to change in biological, mechanical as well as physicochemical properties of a biodegradable biomaterial. [fig1]

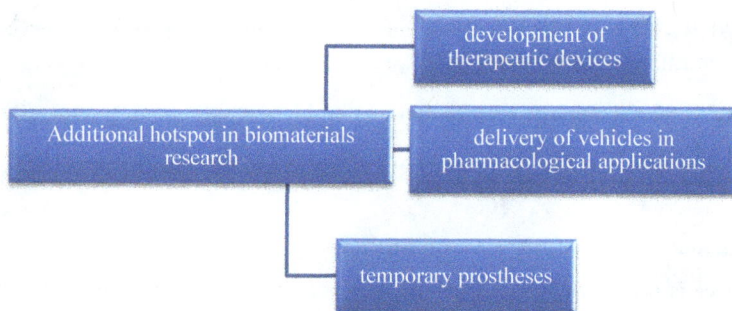

Fig 1. Biomaterials and their applications.

Some renowned applications of polymeric biomaterials are as follows:

1) Large implants

2) Small implants

3) Plain membranes

4) Multifilament meshes [8].

Materials Research Forum LLC

https://doi.org/10.21741/9781644901892-2

2. Advanced polymeric material

The various polymers used in medicinal solicitation are:

- polyethylene (PE)
- polyethylene terephthalate (commonly known as PET)
- polyamide (PA)
- polyurethane (PU)
- polytetrafluoroethylene (PTFE) and polymethylmethacrylate (PMMA)
- synthetic rubber and polystyrene
- polylactic acid (PLA)
- polyetheretherketone
- polyglycolide (PGA) [9]

Polymers recycled & detached into typical polymeric materials such as rubber, cellulose, etc. while artificial in addition to semi-synthetic can be segmented into sustainable or non-biodegradable [10].

Cross-linked type of polymeric materials also have many medicinal applications [11, 12].

Man-made and natural biodegradable polymers vitiate through the body over hydrolysis or enzymes singly. These polymers have been used as medical supplies too due to the specified reason. Synthetic polymers are chiefly polymers having poly(α-hydroxyacid) as one of the units. Natural polymers contain proteins like collagen, albumin, gelatin and polysaccharides. On the other hand, the category of water soluble includes polymers carboxymethyl cellulose which is a natural polymer in first place and poly ethylene glycol (PEG).[13].

- ❖ POLYETHENE have four alternatives [Fig 2]

- ❖ UHMWPE
- It is primarily used on sliding planes of non-natural joints. It is categorized by high influence force, decent biocompatibility as well as chemical constancy.
- Surgical Cables of UHMWPE used in bone rupture, disc substitute for spinal repair, high power orthopedic sutures for soft tissue refurbish, heart valves.

- ❖ PMMA
- Bone bolster for transfer of weight among the insert and the fillet.
- ophthalmic uses(contact lenses).

- ❖ Natural Rubber latex lines its use as an establishment in channeled bone regeneration along with soft tissue healing [14,15].

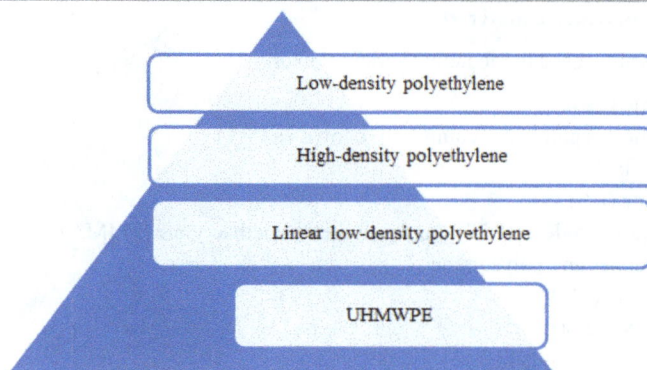

Fig 2 classification of polyethylene

AJ Festas et al (16) shown in his paper the four main kinds of biomaterials are:
The four main kinds of biomaterials are used by AJ Festas et al in his paper(16)

- metal
- polymer
- ceramic
- composites

- ❖ PGA, PLLA poly(L-lactic acid) and PDLA poly(D-lactic acid) are available in various shapes and are therefore marked for the use in orthopedics (in the form of solid body) and drug eluting coating (as meshes).As their initial monomers and products are found to be physiological metabolites, therefore, they are considered for clinical application. The acidic nature of degradation products permits their use in limited amount [17,18].
- ❖ Polyurethanes are investigational polymers for nerve guidance conduits [19-21].

3.   Thermoplastics

Ploy vinyl Acetate (PVC) is the utmost commonly used in medical devices. According to market estimates around one fourth of total plastic medical products are formed by polyvinylchloride.

The main reasons:

- Its low cost
- ease of processing
- The ability to get altered for various other uses
- good transparency
- flexibility
- It can withstand wide range of temperatures be it high or low [22].

Poly ether ether ketone (PEEK)

They are also identified as polyaryl ketones. It is a thermoplastic have Young's' modulus of pliability as 3.6 GPa and tensile strength as 170 MPa. PEEK's melting point is 350°C and is partially crystalline. A polymer having the property of "elasticity" is known as elastomer. It has notably low young's' modulus and comparatively excessive revenue [23].

Poly Tetra Fluro Ehelene (PTFE) is a type of homo polymer (a polymer made up of a single monomer). The advanced Fluro polymer manufacturers preserve the astonishing chemical, thermal and low-friction properties of safe PTFE resin while offering improvements like:

- Weldability
- Improved permeation resistance
- Smoother, less permeable surfaces
- Better high-voltage insulation and Less creep

Sterilization of PTFE can be done by Exposure to high-energy radiation resulting in degradation of PTFE by breaking down the molecules and hence, decreasing its molecular weight which eventually results in a noticeable decline in melts viscosity [24].

Applications and uses of PTFE comprise of:

- fittings
- valves
- tubing
- pumps
- Remarkable developments in the processing of homo fluoro polymers [24]

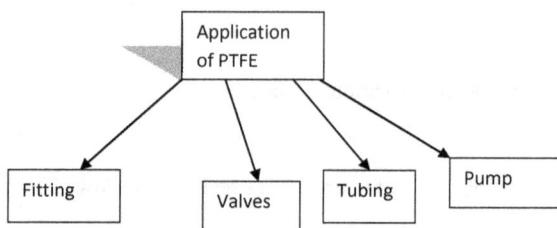

4. Resin

Resins are generally viscous substances of plant or synthetic origin can be typically modified into inflexible polymers by curing process. Resins are now often made artificially by using additives because natural resins are fusible, flammable and soluble in liquids for which they are not suitable for some applications.

Polypropylene resins can be sterilized, thus are commonly incorporated with medical equipment. Due to low density, heat and chemical resistance such resins are used for electronic constituents, toys, pipe, coatings and fibers [25].

These are the most broadly used resin for innovative composite structures and are modified further. Polypropylene resins are formulated to achieve better toughness, high-temperature capabilities, adjustable moisture absorption features and viscosity. As for example different degree of toughness in polymer is achieved by curative resins by controlling the remedial temperature, curing time and method of curing process to achieve cross-linking. Few resins can be cured at normal temperature usually. Maximum resins are alleviated at raised temperature at an autoclave. High range of operating temperature can be achieved by curing resins at elevated temperature which gives increased and greater shear strength. Other curing technologies for very large structures include X-ray cure, UV cure and E-beam cure. The cost of amenities to cure and decrease the complete costs for large structures has compelled the development of these substitute approaches. A resin must be selected to apt the requirements but eventually no solitary resin system can fit all desires. So far several researches have been done on resin to fulfill most of the requirements in polymer science [26].

Vinyl ester resins with unsaturated sites only at the terminal position are the kind of unsaturated polyester resins. Vinyl ester resin is also a class of acrylate functional resin and urethane resins. Such kind of resins is related to epoxy resins. These resins are mostly used for their brilliant mechanical characteristic, hydrolytic resistance and their low styrene emission [27].

Applications of Polymers in Surgery
Materials Research Forum LLC
Materials Research Foundations **123** (2022) 30-52
https://doi.org/10.21741/9781644901892-2

Polymers and their uses in medicinal devices [28].

Table 1. Polymers and their uses.

POLYMERS	USES
PVC	In surgical device, blood purification devices, catheter.
PE	Making bone implants,pouch, containers having flexibility.
PP	Disposable syringesand artificial vascular grafts.
PMMA	Building pumps reused to pump blood and blood purification
PS	Creating filter devices and flasks for tissue culture.
PTFE	In making of Catheter as well as artificial vascular grafts
PU	Generating films incorporated for packaging and molds
PES	Catheters and Lumen tubing
PEEK	Dentistry uses and products
PC	Crystal clear glassy medical devices and valves connectors

PLA polymers mark their use in biomedical as well as pharmaceutical activities. Also, they act as vehicles for controlled drug delivery [29–31].

The internal, intracutaneous closures as well as abdominal surgeries use PGA. The guarantee of holding power over the acute wound healing time period is assured by its great primary tensile strength [32].

❖ The polyethylene and polypropylene matrix composites and their outstanding properties (Shown in table 2)

• low

  ✓ density

  ✓ toxicity

  ✓ moisture absorption

  ✓ gas permeability

  ✓ flammability

- high
  - ✓ transparency
  - ✓ mechanical strength
  - ✓ impact properties
  - ✓ chemical resistance
  - ✓ heat resistance
  - ✓ electrical insulation

They are useful in the pharmaceutical, medical device and diagnostics fields (A.Gopanna et al 2019) [33].

Table 2. Outstanding properties polyethylene & polypropylene matrix composites.

Se no	Properties	Values High/Low
1	Density	Low
2	Mechanical strength	High
3	Moisture Absorption	Low
4	Chemical resistance	High
5	Toxicity	Low
6	Heat Resistance	High
7	Chemical Insulation	High
8	Gas permeability	Low
9	Electrical Insulation	High
10	Impact Properties	High

❖ Eth arylate-based polymers

They have their use in medical field (B. Pomes et al. 2019) [34]

There are several other polymers used for therapeutic device applications in addition to those itemized in this particular chapter. Also, lots of other ways to sterilize are there.

5. Biodegradable polymer in medical devices

There are lots of disadvantages for metals like it's toxic, corrosive properties and similarly bio ceramics are also less interesting because of difficulties in fabrication, low strength in tension and torsion. Polymers have been researched to meet the requirements of the biomedical materials as an alternative of materials composed of metals, alloys and ceramics. Among these materials polymers are versatile in structure, constitution and properties. Due to flexibility in chemistry, polymers possess significant potential giving rise to great diversity in mechanical and physical property [34]. Biodegradable polymers are of great interest as they break down into smaller chemical entities by the cleavage of sensitive bonds hydrolytically or enzymatically and are finally, excreted from the body or resorbed after serving their intended purpose without removal or surgical revision.

Biocompatibility refers to the material should be compatible with living system without producing toxic or immunological rejection after performing its function when exposed to the body in order to direct specific medical treatment (few medical devices, drug delivery, implants and so on) [35].

In the present era the use of natural or synthetic biomaterials have added a great achievement in the medical science due to its biocompatibility with the biological system. Biodegradable polymers are used to synthesize the biomaterials [36]. The ideal characteristics of biodegradable polymers are:

- They should be rich in tenable mechanical properties.
- They must be natural endogenous by-products. The degraded product must be harmless, metabolized and cleared easily.
- The rate of degradation supposed to be as good as with the rate of curing and regeneration.
- They should not exhibit immune reactions.
- They should be simple to synthesize and significantly stable to carry on the fabrication process [35,37,38].

Since more than 50 years, biodegradable polymeric materials have been used for many more applications such as wound dressings, surgical sutures, restricted drug release, regeneration of tissue, immobilization of enzyme, scaffold for tissue engineering, coatings, adhesives, medical embedded devices, artificial limbs and many more.

Applications of biodegradable polymers also include bone plates, bone screws, contraceptive reservoirs, small embedded devices such as staples, sutures, drug releasing vehicles and so on. Such polymers are categorized as natural polymers and synthetic polymers.

The different polymers from the animals or plants are recognized for use as biomaterials. Most of the polymers from nature degrade enzymatically. The use of such natural materials is advantageous for implants due to their similarity to the materials familiar to the body. Natural polymers were the first biodegradable biomaterials used in clinics. Natural materials do not usually exhibit the toxicity which is often experienced by synthetic materials. Again, their protein binding sites and some biochemical signals may help in tissue healing. However, natural materials can have the troubles of immunogenicity. Several limits are there in the fabrication of natural materials of different sizes and shapes for implants due to their tendency to denature and or decompose at temperatures below their melting points. Whereas synthetic polymers have better mechanical properties because of their various chemical manipulation for being able to be produced on a large scale. This group of polymers included polyamide (PA), polyethylene (PE), polyurethane (PU) and others. Disadvantages are there also, like, they are not entirely biocompatible and as a result can cause inflammatory response on the part of the patient, needs care in their use to eliminate this undesirable effect [39]

The remarkable properties of biomedical polymers are specially mentioned because of their light weight, easy to process, better shelf life, less toxicity in vivo and easy to sterilized which are suitable for various applications [40]

In medical field, a large variety of polymers can be found. In this chapter, we emphasize on different biodegradable polymers used in biomedical as well as implantable devices. Materials of medical device and its design should be biologically compatible. [41] The biocompatibility of a polymeric biomaterial is directly influenced by its chemical composition, solubility, molecular weight, its geometry and structure in the case of implants. The biocompatibility varies with time and it is also dependent on the responses of host. The degree of crystallinity is also very crucial for the selection of a polymer in medical applications. Amorphous, crystalline &semi-crystalline are the main three physical states exist in polymeric materials. [42]

The general various fields in which polymers are commonly used: Medicine (syringes, contact lenses, orthopedic uses, membranes, dentistry, cardio vascular equipment etc.

Materials Research Forum LLC
https://doi.org/10.21741/9781644901892-2

The several medicinal Applications of polymeric materials shown below.[39,43,44, ,45-52]

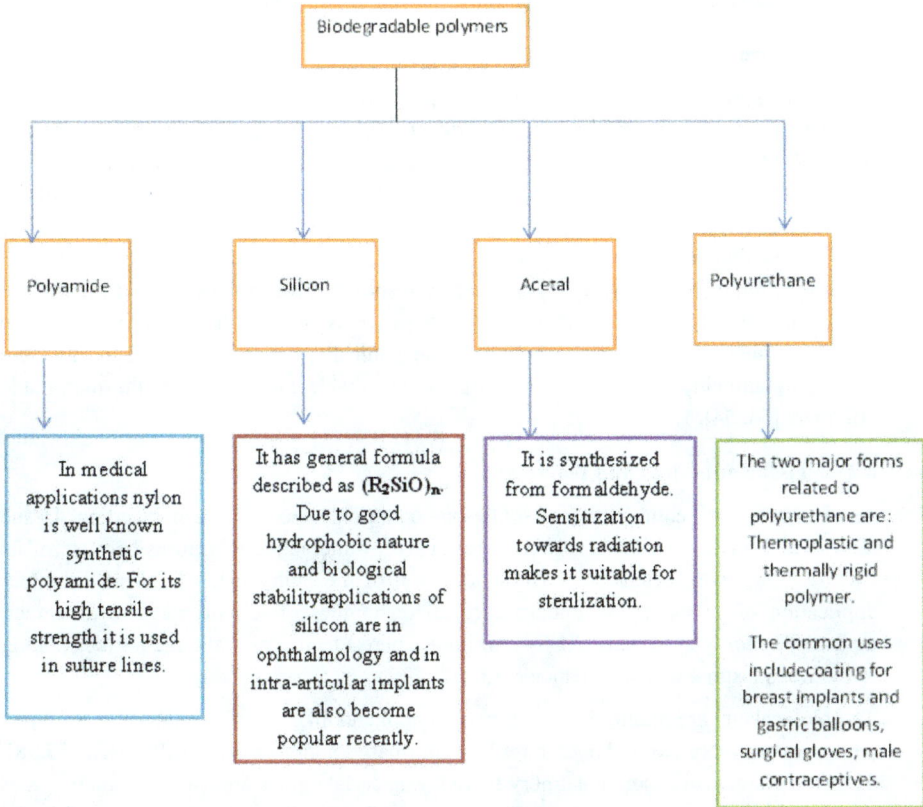

Fig. 3 Biodegradable polymers & their applications.

6.   Polymers in surgery

Synthetic as well as natural biodegradable polymeric substances are used as surgical materials which degrade within the body either through hydrolysis or by the action of enzymes. On the other hand, there are so many uses of water soluble polymeric substances that are not degradable but are absorbable in the body like natural carboxy methyl cellulose and synthetic poly (ethylene glycol) (PEG).

This chapter focuses on polymeric medical devices which are important for surgical operation such as bone fixatives and bio absorbable polymers functional as sealants/adhesives/hemostatic agents, sutures and anti-adhesive barriers.

6.1   Suture materials

In general surgery staples and suture materials are a sphere of polymers. During the selection of suture materials, important factors like friction to tissue, tensile strength, degradability and firmness of knots are to be considered. The functional degradable bio suture materials are collagen based and catgut and non-degradable biopolymers are silk or cellulose(cotton). PGA, PDS, Vicryl, poliglecaprone 25 (Monocryl) are the examples of resorbable polymeric materials where nylon, polyethylene, Prolene, polybutester, polyester and Poly vinyl idenfluorid (PVDF) are the non-resorbable. Resorbable suture material like peritoneum usually used for fast healing tissue, whereas non-resorbable materials are used to heal the tissue like skin or tendons with high mechanical coverage at slower rate. The more intense inflammation is the drawback of non-resorbable materials over the man-made polymers [53, 54].

6.2   Tissue adhesives and sealants

Adhesives with the combination of sutures are commonly used to prevent bleeding. Tissue adhesives with lower adhesion strength are used as alternative of sutures because of its excellent power to form a priori firm occlusion of the injury [55,56]. There are wider application of adhesives in modern surgical techniques of robotic surgical treatment, laparoscopy for body-organs like liver or lung. There a technical challenge for the sticking to the wet substrate which demands extra attention.

The bio-sealants are mainly based on two components like fibrin and thrombi. So many adhesives are there like collagen based, gelatin based and polysaccharide based [57,58]. To avoid stiches, in cosmetic surgery there Cyanacrylate glues are useful synthetic glues in superficial wounds. These synthetic glues are stronger than the fibrin glues. To treat severe wounds in thoracic surgery, photo polymerized PEG-based hydro gels are used successfully. In ophthalmic surgery dendrimers having reactive end groups are applied. The fully degradable tissue glues are Polyurethanes of polycaprolactonediol (PCL) either with isophoronediisocyanate (IPDI) or with hexa methylene di isocyanate (HDI) [59].

6.3   Surgical meshes

Meshes are loosely woven sheet which supports organs or tissue during a surgery. The two important parameters like mesh size and it's weight are relevant factors for the biological response [60,61]. The non-resorbable polymeric meshes with significant degradation at the

Applications of Polymers in Surgery   Materials Research Forum LLC
Materials Research Foundations **123** (2022) 30-52   https://doi.org/10.21741/9781644901892-2

surface and even fragmentation are expanded Poly(propylene(PP), expanded PTFE(ePTFE),Poly vinyliden fluoride(PVDF).

A coupling of metal on UHMWPE is mostly useful in orthopedic surgery, joint prostheses [62]. To improve the mechanical stability by introducing cross linking, generally UHMWPE devices are preferably stored in oxygen containing ambience although it causes ageing of UHMWPE [63,64]. Antioxidant, vitamin E is incorporated to the UHMWPE as a plasticizer to reduce free radicals and to pick up mechanical characteristics [65].

7.   Orthopedic implants

7.1   Bone cements

For more than half of century bone cements serve as the anchor for an artificial joint to the bone to successfully transfer a load coming from the implant. PMMA is widely used for this purpose. Toxicity of monomers may release during the polymerization of PMMA and being an exothermic reaction it causes harm to the tissue.

7.2   Scaffolds for ligament and tendon repair

Mammalian collagen scaffolds from dermis, sub mucosa, small intestine, kidney capsule, pericardium or other tissues are mostly used to connect ligament and tendon [66,67]. Due to having relatively low mechanical stability there is failure in surgery, yet there are some benefits for the dealings with the host tissue, augmentation, remodeling in matrix and cell adhesion. Manmade polymers to repair ligament / tendon are polypropylene and EPTFE, PET/Dacron, nylon providing greater mechanical strength as compared to the natural scaffolds.

7.3   Vascular grafts

EPTFE is the leading material in aneurysm surgery, a hemodialysis access, for bypass surgery [68,69].

8.   Polymeric heart valves

Limited studies are done for polymeric prosthetic heart valves [69,70]. Only two different kinds of fabricated heart valves, either bioprosthetic or impulsive valves are there. For permanent anticoagulation of the patient the impulsive valves show improved extended durability as compare to the bio prosthetic valves [72]. Reduced clotting property of blood, good flexibility and conflict to calcification / degradation have been achieved by thermoplastic polymeric material [73]. The polymeric valves are mainly used in cardiac assisted devices temporarily because their blood coagulation and degradation of valves is a serious issue [73].

9.   Plastic, reconstructive and cosmetic surgery

To correct contour deficiencies tissue augmentation is an only certain sphere of plastic surgery. In chin and malar cosmetic surgery only cross linked silicone elastomer are used by means of contour augmentation on bone and soft tissues. Silicone elastomer is typically used for mammoplasty. HDPE with Medpor is normally used in restoration of nose and reconstruction of ear [74]. Vascular and fibrous tissue delivers accumulation and fixation to the instill [75]. Also, EPTFE is applied as augmenting substance in face [76].

10. Ophthalmology

10.1 Contact lenses

Contact lenses are the biomaterials popularly used on the human eye. Earlier lenses made up of rigid PMMA causes hardness and oxygen impermeability of a contact lens which are harmful to the cornea epithelial cells [77,78]. The material should exhibit hydrophilic nature to maintain the tear film's normal hydration, also should be capable of refreshing tear proteins and lipid deposition. When silicon acrylates came into existence, it gave away for the creation of contact lenses which are permeable to rigid gas. Siloxane comprising hydrogels introduces a great achievement to produce flexible contact lenses with oxygen permeability [79]. Drug release systems quote one major use of hydrogel contact lenses [80]

10.2 Intraocular lenses

Intraocular lenses (IOLs), an implanted polymer devices is most commonly used after cataract surgery in ophthalmology. They are popularly made up of PMMA having outstanding biocompatibility for this application. For the stiffness of these lenses, they demand large incisions in implantation. So as an alternative silicone, copolymers of acrylate, foldable hydrophobic acrylates and methacrylate mixtures of pHEMA with acrylic monomers are used in this purpose [81]. Again biohybrid polymers, containing collagen (Collamer) show good biocompatibility [82, 83]. To protect the retina from UV light and blue-violet light all lenses are prepared by a chromophore [84]. To reduce cell adhesion and opacification, poly (ethylene oxide) which are extremely hydrophobic are incorporated as antifouling coatings as well as fluorinated Omni phobic coatings on the lenses [83]

10.3 Polymer devices in ophthalmology

Silicon is the first option for complicated retinal detachment where vitreous body of the eye is generally detached and desires to replace. If there is any side effect after healing, cross linked hydrogels of Poly- Venyl- acetate, PEG, PVP and poly(acrylamide) are recommended for which the risk of retinal toxicity, glaucoma and cataract progression can be avoided [85,86].

11. Neurosurgery

Peripheral nerve guidance conduits

To repair the peripheral nerve damages nerve guidance conduits are used. The 3D tubular structured conduit provides basic requirements like mechanical stability guiding the axonal sprouting, preventing fibrous tissue in growth. For nerve guidance conduits agarose, keratin, chitosan, silk, poly(hydroxybutyrate) are some of the polymers under investigation [87,88].

11.1 Central nervous system

Because of the high complexity in CNS the regeneration possibilities are more restricted as compared to repair of peripheral nerve. However, there are a variety of approaches by applying hydrogels as scaffold material to redevelop the dopaminergic cells. [89-91]

Conclusion

Presently, lots of polymers are in applications of different medical fields. This wide range of applications are possible because of different modifications in molar mass, crosslinking, co-polymers, degree of crystallization, blends and bioactive surface functionalization. Some mechanical characteristics like stiffness, elasticity and tensile strength are generally major factors to choose a polymer, again biocompatibility aspects should be taken into consideration. Biodegradability is such a superior property of some polymers for which their applications are rising day by day with increasing number of fields because these may vanish after fulfilling their function. These concepts will soon provide promising fields for research and developments which will appear as a great achievement towards the various medical fields.

References

[1] E. Piskin, Biodegradable polymers as biomaterials, J. Biomater.Sci.Polym Ed. 6 (1995) 775–795. https://doi.org/10.1163/156856295X00175

[2] R. Barbucci, Integrated Biomaterials Science. New York: Kluwer Academic/Plenum Publishers, 2002. https://doi.org/10.1007/b112196

[3] R. Langer, J.P. acanti. Tissue engineering Science, New York, 1993. https://doi.org/10.1126/science.8493529

[4] J.P. Vacanti, R.Langer. Tissue engineering: the design and fabrication of living replacement devices for surgical reconstruction and transplantation. Lancet. 354(1999) 32–34. https://doi.org/10.1016/S0140-6736(99)90247-7

[5] T. Dvir, B.P.Timko, D.S.Kohane, R. Langer, Nanotechnological strategies for engineering complex tissues,*Nat Nanotechnol*. 6 (2011) 13–22. https://doi.org/10.1038/nnano.2010.246

[6] Implantable Biomaterials [market report on the Internet.http://www.bccresearch.com/market-research/advanced-materials/implantable-biomaterials-markets-report-avm118a.html. (Accessed February, 2018)

[7] D.S.Kohane, R.Langer, Polymeric Biomaterials in Tissue Engineering, *Pediatr Res*.63(5)(2008)487–491. https://doi.org/10.1203/01.pdr.0000305937.26105.e7

[8] M. Vert, Aliphatic Polyesters: Great Degradable Polymers That Cannot Do Everything. Biomacromolecules. 6 (2005) 538–546. https://doi.org/10.1021/bm0494702

[9] S.Ramakrishna, J.Mayer , E.Wintermantel , et al.Biomedical applications of polymer-composite materials: a Review, 61,(2001) 7789–1224. https://doi.org/10.1016/S0266-3538(00)00241-4

[10] A.Kolk ,J.Handschel ,W. Drescher et al. Current trends and future perspectives of bone substitute materials - From space holders to innovative biomaterials.J.Cranio-Maxillofacial Surg (2012) 40: 706–718. https://doi.org/10.1016/j.jcms.2012.01.002

[11] K. Rezwan, Q.Z. Chen, J.J.Blaker, et al. Biodegradable and bioactive porous polymer Inorganic composite scaffolds for bone tissue engineering. Biomaterials 27, (2006)3413–3431. https://doi.org/10.1016/j.biomaterials.2006.01.039

[12] L.S. Nair and C.T. Laurencin. Biodegradable polymers as biomaterials.ProgPolymSci.32, (2007) 762–798. https://doi.org/10.1016/j.progpolymsci.2007.05.017

[13] D. Lyon, J.Chevalier, L.Gremillard et al. Zirconia as abiomaterial. ComprBiomater 20,(2011)95–108. https://doi.org/10.1016/B978-0-08-055294-1.00017-9

[14] F.A. Borges F,E.D.A. Filho, M.C.R.Miranda MC, et al. Natural rubber latex coated with calcium phosphate for biomedical application. J BiomaterSciPolymEd 26,(2015) 1256–1268. https://doi.org/10.1080/09205063.2015.1086945

Materials Research Forum LLC
https://doi.org/10.21741/9781644901892-2

[15] R.D. Herculano, C.P.Silva,C. Ereno , et al. Natural rubber latex used as drug delivery system in guided bone regeneration (GBR). Mater Res 12(2009) 253–256. https://doi.org/10.1590/S1516-14392009000200023

[16] A.J.Festas, Medical devices biomaterials – A review, J Materials: Design and Applications, Vol. 234(1) 2020, 218–228. https://doi.org/10.1177/1464420719882458

[17] G.O. Hofmann,Biodegradable implant sinorthopaedic surgery – a review on the state-of-the art,Clin.Mater.10(1992)75–80. https://doi.org/10.1016/0267-6605(92)90088-B

[18] S. Kehoe,X.F.Zhang,D.Boyd,FD approved guidance conduits and wraps for peripheral nerve injury: a review of material sand efficacy, Injury 43(2012)553–572. https://doi.org/10.1016/j.injury.2010.12.030

[19] A.R. Nectow ,K. G. Marra, D.L. Kaplan, Biomaterials for the develop-ment of peripheral nerve guidance conduits,TissueEng.PartB:Rev18(2012) 40–50. https://doi.org/10.1089/ten.teb.2011.0240

[20] D. Arslantunali,T.Dursun,D.Yucel,N.Hasirci,V.Hasirci,Peripheral nerve conduits: technologyupdate,Med.Devices(Auckland)7(2014) 405–424. https://doi.org/10.2147/MDER.S59124

[21] A. Faroni,S.A.Mobasseri,P.J.Kingham,A.J.Reid, Peripheral nerve regeneration: experimental strategies and future perspectives ,Adv.DrugDeliv. Rev.82–83 (2015)160–167. https://doi.org/10.1016/j.addr.2014.11.010

[22] A. Bruder, PVC—The Material for Medical Products (translated), Swiss Plastics No. 4, (1999) and Swiss Chem No. 5, (1999)

[23] L.W.McKeen, The effect of temperature and other factors on plastics. Plastics Design Library, William Andrew Publishing; 2008. https://doi.org/10.1016/B978-081551568-5.50004-9

[24] S. Ebnesajjad, Fluoroplastics, volume 1—non melt processible fluoroplastics. William Andrew Publishing/Plastics Design Library; 2000

[25] H. A. Maddah, American Journal of Polymer Science, (2016). 6, 1.

[26] A.A. Vicario, Design and Applications in Comprehensive Composite Materials, 2000.

[27] J. K Fink, Reactive Polymers: Fundamentals and Applications,Third Edition. 2018.

[28] E. Panday, K.Srivastava. S, Gupta, Some biocompatible materials used in medical practices- a review, Int. J. Pharm. Sci. res.2016 DOI: 10.13040/IJPSR.0975-8232.7(7).2748-55

[29] M. Galletti, Organ Replacement by Man Made Devices, J. Cardio thoracic Vascular Anesthesia, 7 (1993) 624–628. https://doi.org/10.1016/1053-0770(93)90327-H

[30] D. Williams, An Introduction to Medical and Dental Materials, Concise Encyclopedia of Medical & Dental Materials, Ed., Pergamon Press and The MIT Press, 1990

[31] A.C. Albersson, I.K. Varma, Recent Developments in Ring Opening Polymerization of Lactones for Biomedical Applications, Biomacromolecules, 4(2003) 1466–1486. https://doi.org/10.1021/bm034247a

[32] S. M.Li Tenon, , H.Garreau, C. Braud, M. Vert, Enzymatic Degradation of Stereocopolymers Derived from L-, DL- and Meso-Lactides, Polym. Degrad. Stab.67(2000) 85–90. https://doi.org/10.1016/S0141-3910(99)00091-9

[33] S.Zhou, Deng, X. Li, W. Jia, L. Liu, Synthesis and Characterization of Biodegradable Low Molecular Weight Aliphatic Polyesters and Their Use in Protein Delivery Systems, J. Appl. Polym. Sci. 91(2004)1848–185. https://doi.org/10.1002/app.13385

[34] B.Pomes·E.Richaud·J-F.Nguyen·Materials for Biomedical EngineeringThermoset and hermoplastic Polymers, Polymethacrylates. (2019), 217-271. https://doi.org/10.1016/B978-0-12-816874-5.00007-4

[35] B. D. Ulery, L. S. Nair, C. T. Laurencin, Biomedical Applications of Biodegradable Polymers,J. Polym. Sci., Part B: Polym. Phys. 49(2011)832–864. https://doi.org/10.1002/polb.22259

[36] R. A. Gross, B. Kalra, Biodegradable polymers for nvironment, Science.297(2002)803–807. https://doi.org/10.1126/science.297.5582.803

[37] R. Chandra, R. Rustgi, Biodegradable polymers, *Prog. Polym. Sci.* 23 (1998)1273–1335. https://doi.org/10.1016/S0079-6700(97)00039-7

[38] L. S. Nair and C. T. Laurencin, Progress in polymer science, *Prog. Polym. Sci.* **32** (2007) 762–798. https://doi.org/10.1016/j.progpolymsci.2007.05.017

[39] A.L.R.Pires, A.C.K.Bierhalz,A.M Moraes, Biomaterials: types, applications, and market. Quim Nova.38 (2015)957–971. https://doi.org/10.5935/0100-4042.20150094

[40] J. C. Middleton, A. J. Tipton, Synthetic biodegradable polymers as orthopedic devices. Biomaterials 21 (2000) 2335-2346. https://doi.org/10.1016/S0142-9612(00)00101-0

[41] Q. Chen, G.A. Thouas, Metallic implant biomaterials. Mater Sci.Eng R Reports 87(2015)1–57. https://doi.org/10.1016/j.mser.2014.10.001

[42] Benavente R. Polı´merosamorfos, semicristalinos, polı´meroscristaleslı´quidos y orientacio´ n, Instituto deCiencia y Tecnologia de Polı´meros, CSIC, Madrid, 1997.

[43] A. Fallis, Bio-materials and prototyping application in medicine. US: Springer, 2008.

[44] S. Mokhtar,S. Ben Abdessalem,F. Sakli, Optimization of textile parameters of plain woven vascularprostheses. J Text Inst. 101(2010) 1095–1105. https://doi.org/10.1080/00405000903363597

[45] T.R Kucklick, The medical device R & D handbook. Boca Raton: CRC Press, Taylor & Francis Group, 2013.

[46] M.F.Maitz, Applications of synthetic polymers in clinical medicine, Biosurf Biotribol 1(2015) 161–176. https://doi.org/10.1016/j.bsbt.2015.08.002

[47] S. Singh, S. Ramakrishna,R. Singh, Material issues inadditive manufacturing: a review. J Manuf Process, 25 (2017) 185–200. https://doi.org/10.1016/j.jmapro.2016.11.006

[48] A.I .Cassady, N.M.Hidzir, L.Grøndahl. Enhancing expanded poly(tetrafluoroethylene) (ePTFE) for biomaterialsapplications, J ApplPolymSci131 (2014). https://doi.org/10.1002/app.40533

[49] M. Arora, E.K.S.Chan, S.Gupta, Polymethylmethacrylate bone cements and additives: a review of the literature. World. J. Orthop. 4 (2013) 67–74. https://doi.org/10.5312/wjo.v4.i2.67

[50] D. Jenke, Evaluation of the chemical compatibility of plastic contact materials and pharmaceutical products; safety considerations related to extractable sandleachables, J.Pharm.Sci.96(2007)2566–2581. https://doi.org/10.1002/jps.20984

[51] R. Vaishya, M.Chauhan,A. Vaish. Bone cement, J ClinOrthop Trauma 4 (2013) 157–163. https://doi.org/10.1016/j.jcot.2013.11.005

[52] R.Ormsby, T. McNally, C. Mitchell, Effect of MWCNT addition on the thermal and rheological properties of polymethyl methacrylate bone cement. Carbon N Y 49 (2011) 2893–2904. https://doi.org/10.1016/j.carbon.2011.02.063

[53] C. Singh, C. Wong, X.Wang, Medical textiles as vascular implants and their success to mimic natural arteries. J.Funct.Biomater.6 (2015) 500–525. https://doi.org/10.3390/jfb6030500

[54] M.F. Maitz, Applications of synthetic polymers in clinical medicine, Biosurface and Biotribology 1 (2015) 161–176. https://doi.org/10.1016/j.bsbt.2015.08.002

[55]J.C.Dumville, P.Coulthard,H.V.Worthington,P.Riley,N.Patel, J. Darcey,M.Esposito,M.vanderElst,O.J.F.vanWaes,Tissue adhesives For closure of surgical incisions,CochraneDatabaseSyst. Rev. 11(2014) 1-135. https://doi.org/10.1002/14651858.CD004287.pub4

[56] L. Sanders, J.Nagatomi, Clinical applications of surgical adhesives and sealants, Crit.Rev.Biomed.Eng.42(2014)271–292. https://doi.org/10.1615/CritRevBiomedEng.2014011676

[57] A. Fu, K.Gwon, M.Kim,G.Tae, J.A.Kornfield, Visible-light-initiated thiol-acrylate photopolymerization of heparin-based hydrogels, Biomacromolecules.16(2015)497–506. https://doi.org/10.1021/bm501543a

[58] L.P. Bre, Y.Zheng, A.P.Pego, W.X.Wang, Taking tissue adhesives to the future:from traditional synthetic to new biomimetic approaches, Biomater. Sci.1(2013)239–253. https://doi.org/10.1039/C2BM00121G

[59] P. Ferreira, A.F.M.Silva, M.I.Pinto, M.H.Gil, Development of a biodegradable bioadhesive containing urethane groups, J.Mater.Sci. Mater. Med.19(2008)111–120. https://doi.org/10.1007/s10856-007-3117-3

[60] U. Klinge, J.K.Park, B.Klosterhalfen, Theidealmesh,Pathobiology 80 (2013)169–175. https://doi.org/10.1159/000348446

[61] U. Klinge, B.Klosterhalfen, Modified classification of surgical meshes for hernia repair based on the analyses of 1,000 explanted meshes, Hernia 16(2012)251–258. https://doi.org/10.1007/s10029-012-0913-6

[62] M.J. Kasser, Regulation of UHMWPE biomaterial sintotalhip arthroplasty,J.Biomed.Mater.Res.B:Appl.Biomater.101B(2013) 400–406. https://doi.org/10.1002/jbm.b.32809

[63] M. Slouf, H.Synkova, J.Baldrian, A.Marek, J.Kovarova, P.Schmidt, H. Dorschner, M.Stephan, U.Gohs, Structural changes of UHMWPE after e-beam irradiation and Thermal treatment, J.Biomed.Mater.Res. B: Appl.Biomater.85B(2008)240–251. https://doi.org/10.1002/jbm.b.30942

[64] I. Urriés,F.J.Medel,R.Ríos,E.Gómez-Barrena,J.A.Puértolas, Comparative cyclic stress– strain and fatigue resistance behavior of electron-beam-and gamma-irradiated ultra high molecular weight polyethylene, J.Biomed.Mater.Res70B(2004)152–160. https://doi.org/10.1002/jbm.b.30033

[65] F. Reno, M.Cannas, UHMWP E and vitamin E bioactivity: an emerging perspective, Biomaterials.27(2006)3039–3043. https://doi.org/10.1016/j.biomaterials.2006.01.016

[66] U.G. Longo, A.Lamberti, N.Maffulli, V.Denaro, T endon augmentation grafts:a systematic review,Br.Med.Bull.94(2010)165–188. https://doi.org/10.1093/bmb/ldp051

[67] J. Chen, J.Xu, A.Wang, M.Zheng, Scaffolds for tendon and ligament repair: review of the efficacy of commercial products, Expert.Rev.Med. Device 6(2009)61–73. https://doi.org/10.1586/17434440.6.1.61

[68] R.Y.Kannan,H.J.Salacinski,P.E.Butler,G.Hamilton,A.M.Seifalian, Current statusofprosthetic bypass grafts:areview,J.Biomed.Mater. Res. BAppl.Biomater.74B(2005)570–581. https://doi.org/10.1002/jbm.b.30247

[69] L. Berardinelli, Grafts and graft materials as vascular substitutes for haemodialysis access construction, Eur.J.Vasc.Endovasc.Surg.32 (2006) 203–211. https://doi.org/10.1016/j.ejvs.2006.01.001

[70] S.H. Daebritz, J. S. Sachweh, B. Hermanns, B.Fausten, A.Franke, J. Groetzner, B.Klosterhalfen, B.J. Messmer, Introduction of a flexible polymeric heart valve prosthesis with special design formitral position, Circulation108(2003)134–139. https://doi.org/10.1161/01.cir.0000087655.41288.dc

[71] B. Rahmani, S.Tzamtzis, H.Ghanbari, G.Burriesci, A.M.Seifalian, Manufacturing and hydrodynamic assessment of an ovelaortic valve made of a new nanocomposite polymer, J.Biomech.45(2012)1205–1211. https://doi.org/10.1016/j.jbiomech.2012.01.046

[72] G. Huang, S.H. Rahimtoola, Prosthetic heart valve, Circulation 123 (2011) 2602–2605. https://doi.org/10.1161/CIRCULATIONAHA.110.979518

[73] D. Bezuidenhout, D.F. Williams, P. Zilla, Polymeric heart valves for surgical implantation, catheter-based technologies and heart assist devices, Biomaterials 36 (2015) 6–25. https://doi.org/10.1016/j.biomaterials.2014.09.013

[74] I. Niechajev, Facial reconstruction using porous high-density polyethylene (Medpor): long-term results, Aesth. Plast. Surg. 36 (2012) 917–927. https://doi.org/10.1007/s00266-012-9911-4

[75] A.S. Breitbart, V.J. Ablaza, Implant materials, in: C.H. Thorne (Ed.), Grabb and Smith's Plastic Surgery, Lippincott Williams & Wilkins, Philadelphia, PA, 2007, pp. 58–65

[76] K.P. Redbord, C.W. Hanke, Expanded polytetraflfluoroethylene implants for soft-tissue augmentation: 5-year follow-up and literature review, Dermatol. Surg. 34 (2008) 735–743. https://doi.org/10.1097/00042728-200806000-00001

[77] D.M. Harvitt, J.A. Bonanno, Re-evaluation of the oxygen diffusion model for predicting minimum contact lens Dk/t values needed to avoid corneal anoxia, Optom. Vis. Sci. 76 (1999) 712–719. https://doi.org/10.1097/00006324-199910000-00023

[78] P.C. Nicolson, J. Vogt, Soft contact lens polymers: an evolution, Biomaterials 22 (2001) 3273–3283. https://doi.org/10.1016/S0142-9612(01)00165-X

[79] S. Kirchhof, A.M. Goepferich, F.P. Brandl, Hydrogels in ophthalmic applications, Eur. J. Pharm. Biopharm. (2015)http://dx.doi.org/10.1016/ j.ejpb.2015.05.016

[80] R. Bellucci, An introduction to intraocular lenses: material, optics, haptics, design and aberration, in: J.L. Güell (Ed.), Cataract, Karger, Basel, 2013, pp. 38–55. https://doi.org/10.1159/000350902

[81] D. Bozukova, C. Pagnoulle, R. Jerome, C. Jerome, Polymers in modern ophthalmic implants – historical background and recent advances, Mater. Sci. Eng. R. Rep. 69 (2010) 63–83. https://doi.org/10.1016/j.mser.2010.05.002

[82] N. Kara, R.F. Espindola, B.A.F. Gomes, B. Ventura, D. Smadja, M. R. Santhiago, Effects of blue light-fifiltering intraocular lenses on the macula, contrast sensitivity, and color vision after a long-term followup, J. Cataract Refract. Surg. 37 (2011) 2115–2119. https://doi.org/10.1016/j.jcrs.2011.06.024

[83] W.J. Foster, Vitreous substitutes, Expert Rev. Ophthalmol 3 (2008) 211–218. https://doi.org/10.1586/17469899.3.2.211

[84] Q.Y. Gao, Y. Fu, Y.N. Hui, Vitreous substitutes: challenges and directions, Int. J. Ophthamol. 8 (2015) 437–440.

[85] A.R. Nectow, K.G. Marra, D.L. Kaplan, Biomaterials for the development of peripheral nerve guidance conduits, Tissue Eng. Part B: Rev 18 (2012) 40–50. https://doi.org/10.1089/ten.teb.2011.0240

[86] D. Arslantunali, T. Dursun, D. Yucel, N. Hasirci, V. Hasirci, Peripheral nerve conduits: technology update, Med. Devices (Auckland) 7 (2014) 405–424. https://doi.org/10.2147/MDER.S59124

[87] A. Faroni, S.A. Mobasseri, P.J. Kingham, A.J. Reid, Peripheral nerve regeneration: experimental strategies and future perspectives, Adv. Drug Deliv. Rev. 82–83 (2015) 160–167. https://doi.org/10.1016/j.addr.2014.11.010

[88] A. Hermann, M. Gerlach, J. Schwarz, A. Storch, Neurorestoration in Parkinson's disease by cell replacement and endogenous regeneration, Expert Opin. Biol. Ther. 4 (2004) 131–143. https://doi.org/10.1517/14712598.4.2.131

[89] U. Freudenberg, A. Hermann, P.B. Welzel, K. Stirl, S.C. Schwarz, M. Grimmer, A. Zieris, W. Panyanuwat, S. Zschoche, D. Meinhold, A. Storch, C. Werner, A star-PEG-heparin hydrogel platform to aid cell replacement therapies for neurodegenerative diseases, Biomaterials 30 (2009) 5049–5060. https://doi.org/10.1016/j.biomaterials.2009.06.002

[90] K.S. Kang, S.I. Lee, J.M. Hong, J.W. Lee, H.Y. Cho, J.H. Son, S. H. Paek, D.W. Cho, Hybrid scaffold composed of hydrogel/3D-framework and its application as a dopamine delivery system, J. Control. Release 175 (2014) 10–16. https://doi.org/10.1016/j.jconrel.2013.12.002

[91] B. Newland, H. Newland, C. Werner, A. Rosser, W.X. Wang, Prospects for polymer therapeutics in Parkinson's disease and other neurodegenerative disorders, Prog. Polym. Sci. 44 (2015) 79–112. https://doi.org/10.1016/j.progpolymsci.2014.12.002

Applications of Polymers in Surgery
Materials Research Foundations **123** (2022) 53-91

Materials Research Forum LLC
https://doi.org/10.21741/9781644901892-3

Chapter 3

Polymers in Tissue Engineering

Z.A. Abdul Hamid[1,a,*], M.M.S. Mohd Sabee[1,b], N.N.F. Nik Md Noordin Kahar[1,c],
N.N.I. Ahmad Tajuddin[1,d]

[1]Biomaterials Research Niche Group, School of Materials and Mineral Resources Engineering,
Universiti Sains Malaysia, Engineering Campus, 14300 Nibong Tebal, Pulau Pinang, Malaysia

[a*] srzuratulain@usm.my, [b]meersaddiq@gmail.com, [c] niknur.farisha8@gmail.com,
[d]nblizzah@gmail.com

Abstract

This chapter presented the numerous types of polymer biomaterial scaffolds for tissue engineering. The different types of clinical applications including bone, cartilage, skin, space filling scaffolds, and adipose tissue regeneration are discussed. Their fabrication techniques; conventional and emerging additive manufacturing for fabrication of the 3D scaffolds are also presented. A focus of hydrogel as scaffolds is discussed in detail. Meanwhile, the common interaction between biomaterials and cells are introduced to ensure the polymer biomaterial scaffolds developed are biocompatible and non-toxic so that the rejection after the implantation could be prevented. The discussed information in this chapter is essential to be considered in any development of a scaffold as an implant for tissue engineering.

Keywords

Tissue Engineering, Surgery, Polymer Scaffold, Polymer Implant

Contents

1.   Introduction

Polymer biomaterials implants or scaffolds have been discovered and developed for the implementation in tissue engineering for quite a long time. This is due to the versatility of polymers in terms of processing, functionality, easy modification, biodegradable, biocompatible, and they can be derived from natural or synthetic polymers and custom made according to the desired properties. The polymer biomaterials can suit the medical applications as their properties can be adjusted from flexible to rigid which can fulfil specific part requirements of the biological body. Furthermore, a wide range of polymers have been granted by the US Food and Drug Administration (FDA) including polyethylene glycol (PEG), polymethylmethacrylate (PMMA), poly(lactic acid) (PLA), poly(lactic-co-glycolic acid) (PLGA), polycaprolactone (PCL), silicone and many more. In this chapter, the importance of the tissue engineering field and their approaches toward the development of polymer biomaterials scaffold as an implant for medical surgery is discussed. The different types of clinical applications for tissue engineering, types of polymeric biomaterials, fabrication techniques and focus on hydrogel scaffold for tissue engineering is explored. Further, the general biomaterials-tissue interaction is introduced for a better understanding on the importance role of polymeric biomaterials properties and their fabrication techniques for the success of the tissue engineering scaffolds.

2.   Tissue engineering

The shortage of organ and tissue donors around the globe has caused millions of patients in need having to risk their life waiting for transplantations. The Ontario's Organ and Tissue Donation Agency, Canada reported that in 2004 the number of patients waiting for transplantation was nearly 2000 compared to only 420 available donors (less than 30 % of waiting patients) (Figure 1) [1]. These numbers slightly decreased after 2004; however, the percentage of donors has remained the same. In the year of 2020, the statistic remained where the number of organ donors are lower than the patients waiting list. There are 108,494 patients waiting for the transplantation and only 35,377 patients went for the transplantation surgeries [2]. These numbers show us that we are now facing a crucial problem; a shortage of organs or tissue for transplantation. Due to this demand, tissue engineering research is now focused on these needs, particularly to help in reducing the shortages of organs and tissue by researching and creating artificial biomaterials that can serve temporarily or permanently as a replacement to damaged or diseased organs or tissue.

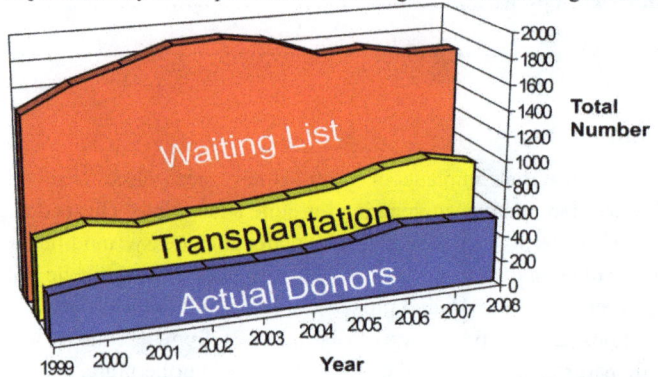

Fig. 1. Statistics of shortage of tissue and organ transplantations reported by the Ontario's Organ and Tissue Donation Agency, Toronto Canada (2009).

3.   Tissue engineering approaches

Tissue engineering is referred to the disciplines which involving biological and physical sciences fundamentals as well as engineering to develop materials for maintaining, restoring or improving damaged organs or tissue [3,4]. There are three major strategies in tissue engineering (Figure 2):

- To restore the specific functions by using cells or cells substitute alone;

- To guide tissue regeneration by using three-dimensional (3D) scaffold alone;

- To grow cells in 3D scaffolds and utilize it for tissue regeneration.

Among these three approaches, the use of 3D scaffolds with the addition of cells has the most potential for regeneration of new tissue [5]. The combination of cells and 3D scaffolds demonstrate the key role for cells homing, proliferation, growth and formation of new tissue or organs. The 3D scaffolds can be functionalized by surface modification or cross-linking with various bioactive molecules such as growth factors for example (i) basic fibroblast growth factor (bFGF)), (ii) fibronectin and (iii) Arg-Gly-Asp (RGD)-peptides in order to improve cellular homing and proliferation [6-10].

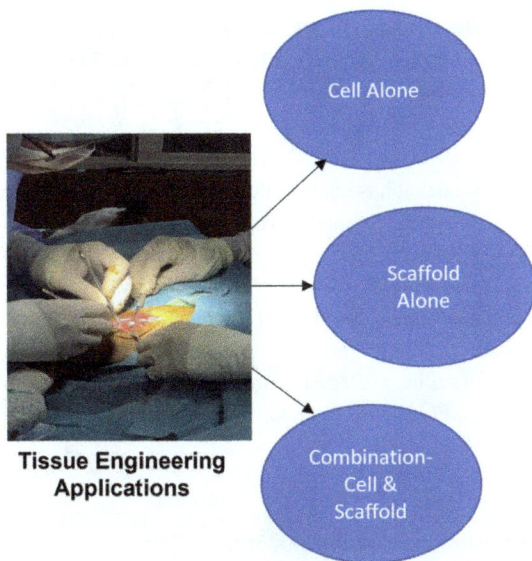

Fig. 2. Tissue Engineering Approaches.

3. Clinical applications of biomaterials in tissue engineering

In tissue engineering, the development of biomaterials is important especially scaffold fabrication. To fulfil the intended applications, the properties of the 3D scaffolds should imitate their targeted site. Each specific site requires specific properties for example the different cells, the tissue functions, mechanical properties requirements, and the explicit environment. It could diverge from hard-tissue, such as bone healing, to soft-tissue, such as skin, bladder, breast, and cartilage restoration [11-19].

3.1 Bone

Generally, the scaffolds designed for this application need to have enough strength for load bearing sites, yet not so strong that they will cause stress-shielding problems. Interconnected pores within the scaffolds play an important role to promote the osteoblasts to form new bones. Among the favoured materials used for bone replacement is coral (exoskeleton composed of calcium carbonate), which acts as a template. Coral basically has adequate mechanical strength and an interconnected pore structure like spongy bone [20-24]. Hydroxyapatite based scaffolds have also been widely investigated for bone tissue engineering. Recently, polymeric materials such as the aliphatic polyesters (poly(glycolic acid) (PGA), poly(lactic acid) (PLA) and their co-polymer poly(lactic acid-co-glycolic acid) (PLGA)), poly(ethylene glycol) (PEG), polyanhydride as well as poly(propylene fumarate) are explored for bone repair [25,26].

3.2   Cartilage

Cartilage refers to tissue that has inadequate ability for self-repair because it is in a form of non-vascular tissue. Generally, chondrocytes are rounded form *in vivo*, however transform to fibroblastic *in vitro*, causing altered gene expression [11]. Thus, ideally the scaffolds should be able to ensure that the rounded morphology of chondrocytes is maintained. Poly(ethylmethacrylate)/tetrahydrofurfuryl methacrylate (PEMA/THFMA) showed the necessary properties to retain the chondrocytes in a rounded shape and have been recommended to be used in cartilage restoration [27]. Materials such as collagen matrices [28], hydrogels [29], PGA and PLA have also been investigated for cartilage tissue engineering [11,20,30-33].

3.3   Skin

In the treatment of burn injuries, the loss of tissue integrity, the loss of massive fluid and microbial penetration are observed. There are three levels of burn injuries; for first degree burns the fabricated skin would be appropriate for full coverage due to the occurrence of non-blistered skin. Whereas, for other levels of burn injuries, deep second-degree and third-degree are able to reduce the reformation of new tissue, as a result, could disrupt their structural integrity in total [11]. Among the FDA approved and commercially available skin regeneration scaffolds are *'Dermagraft'* (Advanced Tissue Sciences) [34] and *'Apligraft'* (Organogenesis) [35]. Even though there are commercial products available for skin regeneration, there are still improvements to be made in tailoring the scaffolds for better formation of new tissue. Hydrogels have potential for skin regeneration because of their unique characteristics and high compatibility with biological environments.

3.4   Space filling scaffolds

There are numerous examples where scaffolds have been used as space filling agents, including bulking, adhesion inhibition and as a physiological glue. For this application, the basic requirement for a scaffold is that it must be able to retain its structural integrity for the specific timepoints [36,37]. Hydrogels are a class of material that have been extensively used as space filling agents. In the urinary incontinence and vesicoureteral reflux treatment, as well as for plastic and reconstructive surgery, the usage of scaffold bulking materials is preferred [36].

Among the natural materials investigated as bulking agents are alginate, chitosan and collagen. Porous RGD modified alginate has been effectively implanted into rats and showed an insignificant immune response with limited encapsulation. A good tissue ingrowth was observed too [36,38]. In contrast, similar studies with porous chitosan scaffolds implanted in mice revealed poor tissue ingrowth, being only observed at the edges of the scaffold [36,38,39].

Synthetic hydrogels are suitable candidates for anti-adhesive applications and have been utilized because of their lack of cell adhesion and protein resistant properties. PEG has been utilized to avoid post-operative adhesion and to safeguard arteries from intimal thickening following damage [36]. One approach to design hydrogel scaffolds for use as anti-adhesives, is by grafting onto poly L-lysine (PLL) to form a PLL-g-PEG copolymer. Theoretically, one surface of the scaffold will adhere to the tissue and the other side will form a non-adhesive brush-like surface [36].

Compared to anti-adhesion agents, physiological adhesives are basically implemented in surgical protocols to stop small injuries, inhibiting the air or bodily fluids from leaking or to boost the efficiency of wound dressing [36]. Commonly used biological adhesive glues are based on fibrin materials. Currently, hydrogels derived from chitosan and chitin derivatives are being investigated as biological adhesives [36,40,41].

3.5   Adipose tissue regeneration

Generally, soft tissue is important for connecting and supporting other tissues that are surrounding the organs in the human body. This includes fat, tendons, muscles, bones, and blood vessels. Though, in reconstructive and plastic surgery, the reconstruction of soft tissue abnormalities remains a difficulty [42]. The repair of soft tissue defects commonly involved a large tissue void which is mainly composed of adipose tissue (i.e: breast reconstruction) [43]. The huge volume loss of adipose tissue often results in the deformation of normal tissue contour [44]. Several strategies have been used such as the breast reconstruction procedures, injection of collagen, and the use of autologous tissue transfer (i.e: free fat tissue grafts and tissue flaps) [44]. Nevertheless, there are many shortcomings of these strategies; for example, the autologous adipose tissue transfer has not been consistently successful in patients [44]. This is because the transplanted tissue resorbed over time which resulted in reduction of the graft (40-60%) loss or the occurrences of fibrosis at the grafted tissue [42,44]. Furthermore, developed adipose cells make up a major portion of autologous adipose tissue, which are unable to differentiate and proliferate. Therefore, the transplantation of adipose tissue on its own cannot promote repair of the damaged tissue [42,45-47].

Applications of Polymers in Surgery   Materials Research Forum LLC
Materials Research Foundations **123** (2022) 53-91   https://doi.org/10.21741/9781644901892-3

Currently, immense researches had been conducted in the development of engineered adipose tissue. There are two main strategies to induce adipogenesis [42-44,48,49]. The first strategy is the use of primary adipogenic stem cells or preadipocyte cell lines to induce the regeneration of adipose tissue [49,50]. Among the most explored adipogenic cell lines are Ob17, 3T3-F44a and 3T3-L1 [50-52]. In comparison to primary cells, these cell lines can differentiate quickly and transform into adipocytes in the presence of serum, and their development can be sustained to confluence [50,53]. Bone marrow is an attractive cell which is capable of adipogenic differentiate and a source of mesenchymal stem cells (MSCs) and haematopoetic stem cells [50,54]. Stem cells are favourable because they have the capability to split and replenish themselves over a long duration of time. They also could differentiate into many cell types which quite simple to isolate and expand [49,54-56]. Furthermore, stem cells could also differentiate into adipocytes and endothelial cells when exposed to appropriate stimuli and possibly can develop a vascularised fat graft for reconstructive applications [44,49]. The second strategy is the implantation of scaffolds with the combination of preadipocyte cells that are able to create adipose tissue at the specific location where the construction of adipose tissue is required [43]. To date, numerous types of scaffolds derived from synthetic and natural materials have been investigated to fulfil the need of adipose tissue engineering applications. Ideally, the implanted scaffold should remodel or absorb to give way for seeded preadipocytes to grow, proliferate and ultimately establish into adipose tissue. Several examples of scaffold utilized in adipose tissue engineering are polyester based absorbable (PLLA, PLGA and PGA), hyaluronic acid, collagen, polyethylene glycol-diacrylate (PEGDA) hydrogels and chemically modified alginates [44,50]. Hydrogel scaffolds showed desirable properties for use in the soft tissue engineering applications. This is because their swollen structure that give mechanical characteristics mimic the soft tissue and extracellular matrix [57]. Moreover, hydrogels are hydrophilic in nature and have physiological recognition qualities, which means they can surround cells within their hydrated matrix [58-60]. Alhadlaq and co-workers reported that poly(ethylene glycol) diacrylate (PEGDA) hydrogels encapsulated with bone marrow mesenchymal stem cells have exhibited great prospects to facilitate adipose tissue formation [57]. Differentiation was demonstrated *in vitro* and *in vivo* via subcutaneous athymic mice after 4 weeks of seeding. The PEGDA hydrogel scaffold maintained their structure after 4 weeks *in vivo* and from the later findings, they assured that the PEGDA scaffolds in combination with cells have facilitated the formation of adipose tissue. Another interesting outcome of this work is the scaffold volume retention characteristics was maintained after 4 weeks *in vivo*. Particularly, 100% volume retention was successfully achieved for PEGDE/cells scaffolds, whereas, within the similar system, collagen gels retained 35-65% of their volume [57]. Meanwhile, poly(ethylene glycol)-fibrinogen (PF) also has been utilized as a biocompatible hydrogel for the 3D culture of three types of breast cancer cells. From the studies, PF-based hydrogels enabled 3D culture of breast cancer cells and showed a potential cellular behaviour in consideration of changing matrix properties. The cell proliferation and 3D morphology were successfully observed within the PF hydrogel scaffolds [61].

The use of natural extracellular matrices (i.e: Matrigel^{TM} and Myogel) with addition of appropriate growth factors also found to be promising in inducing adipogenesis *in vitro* and *in vivo* [49,62,63]. Kawaguchi reported that Matrigel enhanced with FGF2 (at a dose of 1μm/mL Matrigel) which subcutaneously injected in mice had induced *de novo* adipogenesis. In another work, Cronin and co- workers have established a novel technique that permits the construction of *de novo* vascularised adipose tissue *in vivo* in murine chamber model [64]. The technique demonstrated angiogenesis and adipogenesis has occurred orchestrated to a vascularised fat flap on the vascular pedicle. Therefore, it is possible that the resulting adipose tissue can be surgically transferred to any defect site in the body. Even though there are many successful studies reported in the adipose tissue engineering field, there are limitations and unpredictable events often occurred during the *in vivo* study using both strategies mentioned above. Therefore, enormous works still have to be done and explored in developing engineered adipose tissues which fulfil the specific requirements of adipose tissue engineering applications.

4. Polymeric biomaterials as scaffolds for tissue engineering applications

Materials commonly utilized in scaffold fabrication include naturally derived and synthetic components. The selection of the material is based on their specific application.

4.1 Naturally derived biomaterials

There are various types of materials that have been employed in the fabrication of scaffolds. Among them are naturally occurring materials such as alginate, hyaluronan, chitosan, polysaccharides, collagen, gelatin and fibrin (Figure 3) [13,65-70]. Naturally derived materials may closely simulate the native tissues however, they require difficult purification procedures and can carry a risk of transmission of diseases [71,72]. They also have disadvantages in clinical use, mainly due to their low mechanical strength, which can cause complications in handling during surgery [71].

4.2 Synthetic biomaterials

Synthetic materials offer advantages over natural materials as various properties can be tailored making them more predictable. Synthetic materials commonly utilized and authorized by FDA are PGA, PLA and their co-polymer PLGA, polycaprolactone (PCL) and polyanhydrides (Figure 4) [25,66,73,74]. However, these materials are very

Applications of Polymers in Surgery   Materials Research Forum LLC
Materials Research Foundations **123** (2022) 53-91   https://doi.org/10.21741/9781644901892-3

hydrophobic in nature, therefore, potentially could cause limited cell attachment upon the scaffold surfaces, and therefore, reduced cell proliferation and growth. Another drawback is the acidity of their decomposition products and slow degradation rate [29,66,75,76]. Although there have been improvements made by modifying the scaffolds' surfaces, the success is very limited.

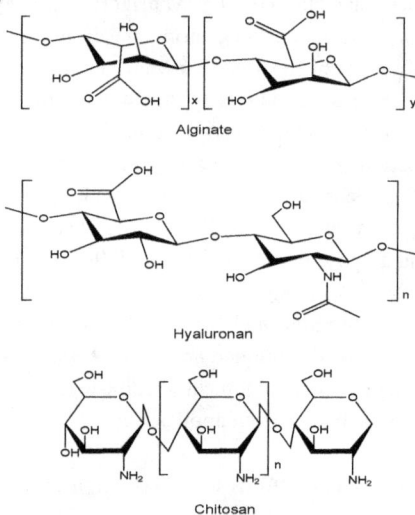

Alginate

Hyaluronan

Chitosan

Fig. 3. Chemical structures of naturally derived materials; alginate, hyaluronan and chitosan.

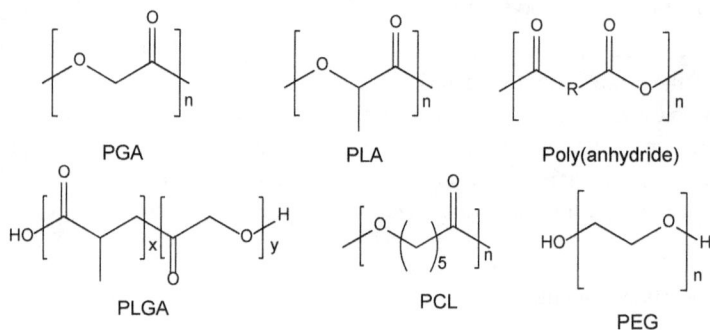

PGA

PLA

Poly(anhydride)

PLGA

PCL

PEG

Fig. 4. Chemical structures of poly(glycolic acid) (PGA), poly(lactic acid) (PLA), poly(anhydride), poly(lactic-co-glycolic acid) (PLGA), poly(caprolactone) PCL and poly(ethylene glycol) (PEG).

Another synthetic polymer, namely poly(ethylene glycol) (PEG), possesses more desirable properties, such as high hydrophilicity and biocompatibility (Figure 4). PEG has been employed in a numerous biomedical applications which includes the preparation of physiologically relevant conjugates, cell membrane fusion induction and biomaterials surface modification, as well as scaffolds for tissue engineering applications [46].

5.   Scaffold fabrication techniques

The structural morphology of the fabricated scaffolds is another important aspect in designing scaffolds for tissue engineering. Ideally, scaffolds must have interconnected and suitable pore sizes for promoting the formation of new tissues. Table 1 shows the advantages and shortcomings of several scaffold fabrication methods utilized in tissue engineering applications.

Table 1. Fabrication techniques of scaffolds for tissue engineering applications [67].

Fabrication Technique	Advantages	Disadvantages
Solvent casting and particulate leaching.	Wide range of pore sizes. Porosity and pore size can be easily controlled Crystallinity can be tailor-made. Highly porous structures.	Limited membrane thickness dependent on scaffold materials. Closed pore and limited interconnectivity. Porogens residue. Difficult to control the internal architecture.
Fibre Bonding.	High porosity.	Small range of polymers. Solvents Residue. Low mechanical strength.
Phase separation.	High porosity. Allows bioactive molecules integration.	Difficult to regulate the internal structure. Strict pore sizes range.
Freeze drying.	Highly porous structure. High porosity and interconnectivity.	Only obtained small pore sizes.
Stereolithography.	Ease of use, suitable for small features characteristics	Choice of polymers (photosensitive and biocompatible) is limited.

		Exposure to laser.
Gas Forming.	Easy processing.	Poor control over internal architecture.
	Porous foam structure.	
		Scaffold too compact.
Rapid prototyping.	High porosity and fully interconnected structure.	Limited choice of polymers.
	High accuracy of pore size and pore shape.	Applicable only for solid structure.
		Residual solvents.
	Able to fabricate scaffolds with complex shaping.	Expensive.

Perhaps the most widely employed scaffold fabrication technique that has been widely used is solvent casting/particulate leaching (SC/PL) [77-82]. This technique is favoured because it involves simple processes and suitable for a wide variety of polymers. SC/PL can produce scaffolds with a controlled pore size and volume. These properties could be controlled by monitoring the size and quantity of initial particulate used to prepare the template. However, the size of the scaffolds is limited by this technique, which has lead to the development of other techniques such as thermally induced phase separation [54,55], fibre bonding [80,83], freeze drying [84], gas forming [85], rapid prototyping, stereolithography [86] and injection moulding [87] (Table 1). Another limitation of this technique is the formation of closed pore morphology that subsequently preventing the migration of cells within the scaffolds and therefore undesirable for cells proliferation and growth.

5.1   Solvent casting and particulate leaching (SC/PL)

The solvent casting and particulate leaching (SC/PL) technique was first introduced by Mikos and co-workers in the early 1990s [83]. SC/PL is a simple technique (Figure 5), which initially involves dissolving the scaffold precursors in a solvent (solvent chosen depending on the solubility of the precursors used).

The precursor mixture is then poured into the prepared template. Various types of porogen can be used in preparing the template, including inorganic salt particles (i.e: sodium chloride [77,81,83,88], ammonium bicarbonate [89]) and organic particles (i.e: paraffin spheres, sugar particles, gelatin spheres) [90]. The scaffolds with a suitable pore size can be regulated by varying the porogen particle sizes. The solvent in the resulting scaffold/gel is then removed via evaporation or dissolution. The templates are leached out using appropriate solvent (i.e: water for sodium chloride and hexane for paraffin spheres) to produce the porous scaffold.

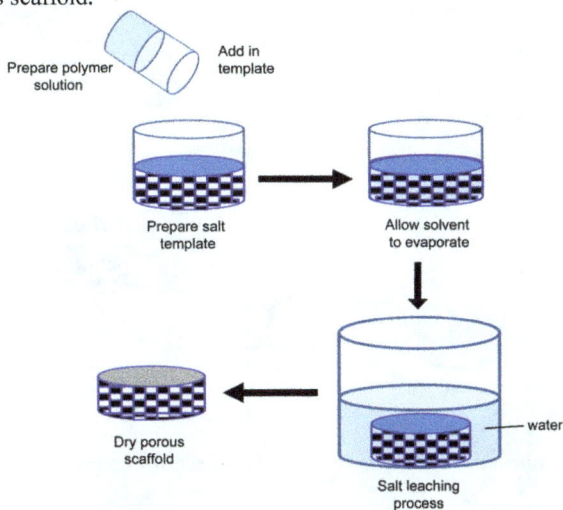

Fig. 5. A schematic of the solvent casting/particulate leaching (SC/PL) technique.

5.2   Thermally induced phase separation (TIPS)

Thermally induced polymer separation (TIPS) is based upon the concept of phase separation, which entails dynamic demixing of uniform polymer-solvent mixture into polymer-poor and polymer-rich phases. This can be done either by exposing the solution to another insoluble solvent or cooling the solution lower than binodal miscibility curve. In the TIPS approach, thermal energy is employed as a driving force to inflict phase separation (Figure 6) [89].

Examples of polymers employed in this approach are poly(L-lactic acid) (PLLA) and poly(lactic acid-co-glycolic acid) (PLGA) [83,89,91-93]. For this method, firstly, the polymers are dissolved in a solvent with a low melting point (i.e: phenol, naphthalene, or 1,4-dioxane) in order to make sure that they easily sublime. The polymer solution is then cooled down below the melting point of the solvent. Whereas, controlled cooling parameters are vital as they will give impact to the morphology of the resulting scaffolds. Subsequently, vacuum drying is conducted for complete solvent removal. Lastly, the polymer-poor phase is eliminated creating a polymer network with high porosity characteristic.

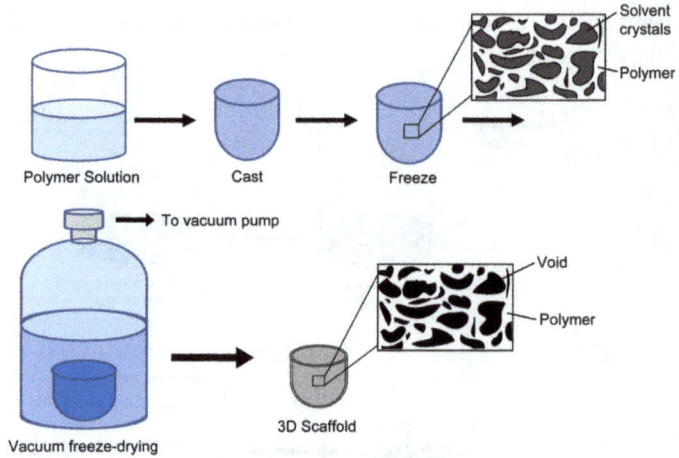

Fig. 6. A schematic of the thermally induced phase separation (TIPS) technique.

5.3   Fibre bonding

PGA fibres used as scaffolds were among the earliest constructs for tissue engineering proposed by Freed and co-workers and Mikos and co-workers [83,88]. Figure 7 shows the technique that was developed by Mikos and co-workers, where the PGA fibers are deep in PLLA solution.

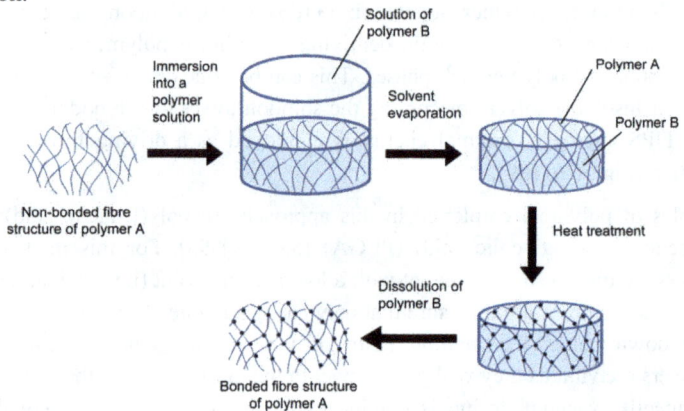

Fig. 7. A schematic showing the fibre bonding technique.

The solvent is left to evaporate to afford PGA fibers incorporated PLLA. Then, the composite is heated over the polymer melting temperature. After heating, PLLA is eliminated via dissolution using dichloromethane. This method produces highly porous foam structure morphology < 81 % and pore size (500 μm). The limitations of this technique include incomplete removal of the solvents and high temperature used during heat treatment, which causes problems if either bioactive agents or cells have to be incorporated during scaffold fabrication process [83].

5.4 Gas forming

The gas forming technique necessitates high-pressure CO_2 gas processing to prepare porous scaffolds. Basically, the volume of gas dissolved in the polymer mixture impacts the pore morphology, including pore diameter and pore structure, of the resultant scaffolds. Secondly, it is also depending on the nucleation rate and type of gas used. Further, the pore morphology is also influenced by diffusion rate of the gas molecules from the polymer mixture into the pore nuclei. Figure 8 shows a schematic of the gas forming technique. The polymer mixture is placed at the top of a cell strainer; while at the bottom of the beaker is a controlled amount of porogen salt ($NaHCO_3$) in chloroform. Hydrochloric acid (HCl) is injected to accelerate the production of CO_2 bubbles. The limitation of this technique is the produced scaffold is too compact and is not viable for cell attachment and growth.

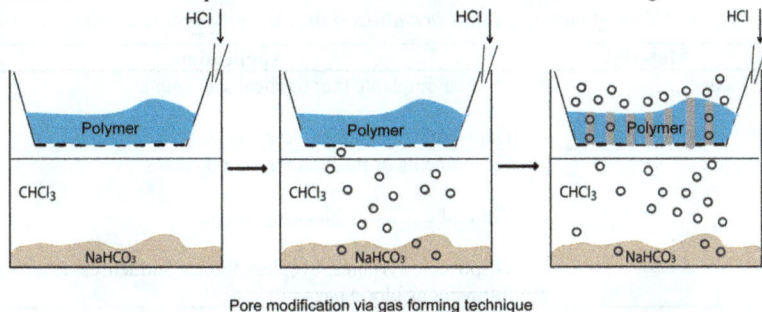

Pore modification via gas forming technique

Fig. 8. A schematic of the gas forming technique.

5.5 Additive manufacturing

Additive Manufacturing (AM) is a 3D printing procedure that uses successive layering sequences to produce objects. Based primarily on input materials, AM is typically classified into several techniques including power-based, solid and liquid [94]. Selective Laser Melting (SLM) and Selective Laser Sintering (SLS) are considered as power-based patterns, whereas Fused Deposition Modeling (FDM) is categorized as solid-based pattern. Direct Ink Writing (DIW), Optical Light Processing (DLP), Stereolithography (SLA), and Inkjet are examples of liquid-based trends [95].

Furthermore, AM exhibits flexibility for the manufacture of any shape, and these emerging innovations are having great potential for the medical therapy as well as the future industrial revolution [96]. In fact, AM creates precise forms of patient-specific software for intricate procedures. In cardiology, special heart valves and scaffolds are used, and they are also useful for pre-operative planning. [97]. In comparison to conventional technologies, this method is able to manufacture surgical equipment at lower expense. Other than that, it also makes one-of-a-kind prosthesis that are simple to modify and blend in with the wearer's appearance [98]. By removing human organs, creating prosthetic organs, surgical devices, and speeding up surgery, AM technology serves to improve human life. In the realm of medical treatment, effective contact and chemicals are essential. This technology is used to produce biomaterials that are suitable for bioprinting. For the medical laboratory, a live cell and synthetic living tissue can be generated. [99]. This makes creating surgical equipment to protect tissues and organs throughout surgery easier. To summarize, broad range of biomaterials are implemented to construct 3D models in the area of medicine. Table 2 lists the types of biomaterials utilized in AM technologies.

Table 2. List of materials that are utilized in AM in the medical industry [100].

Material	Application
Silicone.	Medical implants (i.e: tracheal scaffolds).
	Diagnostic equipment, other surgical instruments.
Plastic.	Components of medical wires and cables.
	Thin tube to aid unblocking arteries.
	Disposable syringes, eyeglass frames and lenses, heart valves, intravenous blood bags.
Metal.	Metals implants to repair a damaged bone.
	Scaffolds, plates, artificial joints, screws and valves.
	Prolonged implantable devices.
Tissue and cell.	Improving the entire organ or damaged tissue.
	Generating desired cell type.
	Human tissue repair products, organs for transplantation.

Poly(lactic acid) (PLA).	Tissue engineering (i.e: bone application).
	Dental extraction for wound treatment, postoperative adhesion prevention, and surgical sutures.
	Scaffolds.
Carbon Fiber.	Medical, limb loss.
Powder.	Pre-operative strategy for complicated surgery.
	Medical student's curriculum.
Nitinol.	Personalization, medical implants.

AM also familiar as 3D printing is basically gaining popularity due to the numerous applications it has in a variety of fields, the most important of which are pharmaceutical and medical science. This adaptable technology has made tangible and innovative breakthroughs in targeted therapy and diagnostics as well as bio-inspired devices, in response to a growing demand for tailored devices in targeted therapy and diagnostics. This technique involves printing one or more materials layer by layer with a 3D printer and then adjusting the shape of each individual layer to form a complex, solid object from a digital model. Figure 9 shows the workflow of additive manufacturing. A detailed workflow is discussed in the next sub section.

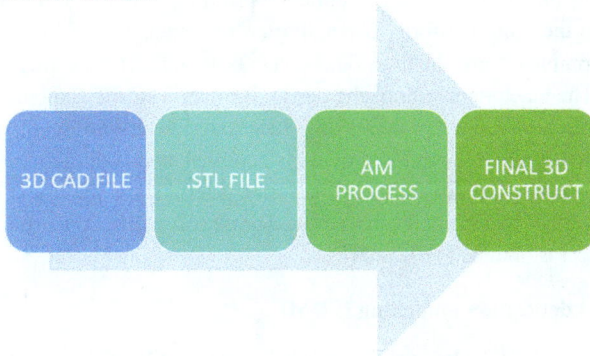

Fig. 9. Work-flow of AM printing process.

5.5.1 3D printing technology

3D printing in medical applications is constantly evolving, and it is believed to have a substantial impact on health care. Current and prospective 3D printing for medical uses include tissue and organ manufacturing, implants and anatomical models, customised prosthetics, as well as pharmaceutical research into medication dosage formulations, administration, and discovery. The use of 3D printing in medicine has several advantages, including the ability to customize and personalize medical supplies, medications, and equipment. Other than that, the productivity will increase as a result of the cost-effectiveness, as well as design and production which will become more progressive. However, despite recent substantial and promising medical discoveries involving 3D printing, there are still enormous scientific and regulatory barriers to overcome, and the most challenging area of research will take a while to develop [101].

5.5.2 The 3D printing technology principles

3D printing is the technique of turning a computer model into a three-dimensional construct of nearly any shape. 3D printing, also known as AM, is a technology that uses an additive process to lay down successive layers of material in various designs. Industrial design, automotive, aerospace, medical, tissue engineering, architecture industries, and even the food business employ 3D printing technology for prototyping and dispersed manufacturing.

The first stage in 3D printing is modelling. Virtual models built in with computer-aided design (CAD) or animation modelling software are "sliced" into digital cross-sections, which the machine utilizes as a guideline for printing. Material or a binding substance is deposited on the build platform, depending on the machine, until the material or binder layering is completed, and the final 3D object is generated. The second phase in 3D printing is printing. The machine interprets the design from an STL file and builds the model from a series of cross-sections by depositing successive layers of liquid, powder, or other materials. Finally, the 3D-printed product is completed according to the model. Some 3D printing protocols allow for the use of several materials and the usage of supports when constructing parts. The supports are removed or dissolved after the print is completed, and the final object is obtained [102, 103].

5.5.3 Fused deposition modelling (FDM)

Compared to SLS printers, FDM printers are far more popular and less expensive. Like an inkjet printer, the FDM printer features a print head. FDM uses a polymer filament as the feed material to build the layer-by-layer 3D printed product. As the print head advances, molten beads of heated plastic are released, producing the item in tiny layers. This process is repeated numerous times, allowing for exact control of the amount and placement of each deposit, allowing each layer to be shaped [104]. The material fuses or connects to the layers below due to being heated during the extrusion process. Each layer of plastic

Applications of Polymers in Surgery
Materials Research Foundations **123** (2022) 53-91

Materials Research Forum LLC
https://doi.org/10.21741/9781644901892-3

solidifies as it cools, eventually becoming the solid object. Leaning on the specificity and expense of the FDM printer, enhanced functionality such as numerous print heads may be implemented. FDM printers are capable of printing on a wide range of materials. Indeed, because FDM printed components are typically made of the similar thermoplastics utilized in conventional injection moulding, they share mechanical, durability, and stability characteristics [105].

The melting and bonding of thermoplastic material with the polymer serving as the printing material is the basis of the FDM production method. The approach is gaining popularity due to the advantages of the thermoplastic material employed, which include long life, easy availability, high toughness, and the capacity of some types to dissolve rapidly in nature, as well as low shaping temperature and reshaping when heated. This approach can now be used to create a variety of parts at home. Due to the simple operating principle and cheap equipment requirements, the costs of devices manufactured using the FDM method are lower than those generated using other AM methods.

The FDM process generated devices known as 3D printers that are ideal for desktop usage as a result of these properties. Because of the low cost and other benefits of both devices and thermoplastic materials, individual users have used the FDM approach. Thermoplastics are polymer materials that can be softened and shaped while still in a solid form at low temperatures and are used in FDM production. After completion of the thermoplastic forming process, the product cools and solidifies, resulting in the desired geometry [106].

5.5.4 Advantages of 3D printing in tissue engineering applications

The advantages of adopting 3D printing in the medical industry are first and foremost customization and personalization. The fundamental benefit of 3D printers in medical areas is the ability to create custom-made medical devices and instruments [107]. 3D printing, for example, can be immensely advantageous to patients and surgeons when it comes to personalizing implants and prosthetics [107]. Furthermore, unique jigs and equipment for utilizing in surgical units can be implemented using 3D printing [103]. Fixtures, implants, and equipment created for surgical purposes can save time, enhance recovery period, and increase operation or implant success [103]. In the future, 3D printing technologies are predicted to enable tailored drug dose forms and release profiles for each patient [101].

Next, the advantage of 3D printing is its greater cost efficiency, as it can make products at a low cost [108]. Conventional manufacturing processes are still more cost-effective for large-scale production, while 3D printing is far more profitable for minor operations [108]. Particularly, it is accurate for minor conventional implants or prosthesis used for dental, spinal, or craniofacial diseases [107]. Custom-printing 3D product is inexpensive, as the first piece costing as cheap as overall [108]. Therefore, this is notably beneficial for

Applications of Polymers in Surgery   Materials Research Forum LLC
Materials Research Foundations **123** (2022) 53-91   https://doi.org/10.21741/9781644901892-3

businesses with modest production numbers, as well as those who create highly complicated parts or products that require frequent adjustments [103].

By reducing the usage of unnecessary materials, 3D printing can help save manufacturing costs. For instance, 10 mg pharmaceutical tablet might be custom-made as a 1-mg product on demand. In addition, some pharmaceuticals could be produced in dose forms to make patient delivery more convenient and cost-effective [101].

Furthermore, one of the pros of 3D printing in the medical industry is increased production. In 3D printing, "fast" denotes as a quality that can be created in a few hours [103]. As a result, 3D printing is far faster than conventional techniques of producing prosthetics and implants, which require forging, milling, and a long lead time. Other features of 3D printing technology, such as resolution, precision, dependability, and repeatability, are also advancing [107].

Last but not least, the democratization of product design and manufacturing is an advantage of 3D printing technology in the medical industry. A growing number of materials are nowadays feasible for 3D printing, apart of their falling prices. This enables more people, especially those in medical disciplines, to develop and build unique products for personal or commercial use by 3D printing and maximize their creativity.

The nature of 3D printing data sets also provides a once-in-a-lifetime chance for scholars to collaborate. Rather than attempting to replicate settings reported in scientific articles, researchers can use open-source databases to get downloadable .stl files [104]. In this way, they can utilize 3D printing in order to manufacture an identical reproduction of medical equipment, allowing for precise design sharing.

6.   Hydrogel as scaffolds for tissue engineering applications

Hydrogels have abundant utilizations in tissue engineering specifically in medical surgery as well as in drug delivery. Currently, hydrogels as scaffolds have emerged as significant replacements for regeneration of new tissues [66]. Tissue engineering scaffolds can be derived from natural and synthetic hydrogel scaffolds. For tissue engineering applications, scaffolds must meet numerous characteristics, despite of materials' type utilized [109]. Therefore, researchers need to investigate various aspects which will affect and contribute to the success of the final product before designing any biomaterial or scaffold.

6.1   Hydrogel definition

A hydrogel is defined as a material that is hydrophilic in nature and can exhibit the characteristic macromolecular structure of a gel. A hydrogel is best described as a continuous 3D network linked together by chemical or physical bonds [110]. They have the aptitude to absorb water many times their own weight and can emulate the physical characteristics of soft tissue [111]. Hydrogel is classified into two categories based on the types of bonds of which they comprise: covalently and physically cross-linked hydrogels (Figure 10). The solid circles in Figure 10(a) represent the *covalently cross-linked points*, while in Figure 10(b) the highlighted area shows a *crystalline region* which holds the network together by physical forces for example hydrogen bonding, van der Waals forces within physical hydrogels. The highlighted area could also represent the junction zones which can be ordered but be non crystalline or they can form micelle junctions in triblock polymers.

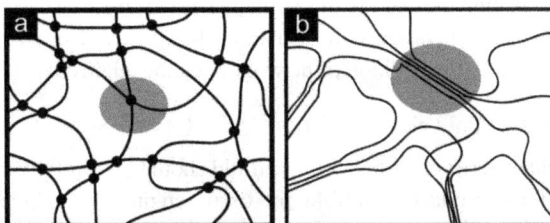

Fig. 10. Structure of (a) covalently and (b) physically cross-linked hydrogels.

6.1.1 Covalently cross-linked hydrogels

In general, covalently cross-linked hydrogel will not dissolve in water or other organic solvents upon heating except of their covalent bonds are broken. Relatively, they will swell in the presence of water or organic solvents. There are two general techniques to prepare covalently cross-linked hydrogels: (i) polymerization of a hydrophilic monomer in the occurrence of a suitable cross-linking agent, or (ii) cross-linking a hydrophilic polymer with an appropriate cross-linking agent. These techniques follow a common reaction mechanism and can be employed in many ways. However, the monomers and cross-linkers used need to have a functionality of more than two or three, respectively, in order to form a 3D network [110].

The solvent used in the cross-linking reaction is chosen with regards to the final application of the hydrogel. In biomedical applications, water is preferred as the solvent because of its non-toxic properties. However, ketones and alcohols, on top of other organic solvents can still be used if they can be easily removed and replaced by water (or biological fluids) in the final hydrogel network.

6.1.2 Physical hydrogels

Physical hydrogels are networks held together by secondary forces, which are influenced by the molecular structure and arrangement of the polymers. Examples of intermolecular forces that contribute to these physical bonds are London dispersion forces, permanent dipoles, hydrogen bonding and ionic interaction [110]. These forces are abundant in nature and contribute immensely to the well-organized structure displayed by biological macromolecules. The physical bonds that hold these types of hydrogels together represent the *ordered state* of the hydrogels (Figure 12(b)). These segments are comprised of molecules that position themselves in a strongly packed structure and in a repeated manner [110]. These types of hydrogels can generally be dissolved in water or solvents upon dissolution, heating or the addition of network breaking additives [110].

6.2 Hydrogel scaffold characteristics

The specific characteristics of hydrogel scaffold should be carefully considered to ensure the success of the produced scaffold for tissue engineering applications. Among the characteristics are, (i) biocompatibility with the physiological surrounding, (ii) acceptable mechanical characteristics based on the targeted site, (iii) suitable pore morphology with specific pore sizes and interconnected pores structures and (iv) biodegradability that is matching with the rate of tissue regeneration [11,36,67,109].

6.2.1 Biocompatibility

The designed scaffolds need to be biocompatible, thus they should be able to be implanted *in vivo* without causing any cell damage or significant scarring, which will alter the cells desired functions. A severe inflammatory response to the hydrogel scaffold can influence the immune response of the body and therefore, introduce problems towards the transplanted cells or *vice versa* [66]. The scaffold materials and its degradation products must be non-toxic to the cells and should favourably interact with cells, promoting cell homing, growth, migration and differentiation to generate new tissues.

6.2.2 Mechanical properties

3D scaffolds in tissue engineering are important in providing the required support for cells homing, proliferation and maintaining their differentiated function as its structure reflects the final shape of the new tissues. For this reason, the scaffolds need to have appropriate strength to support cells and have to be resistant to external forces until the new tissues are formed. Ideally the mechanical function (i.e: ultimate strength, elasticity and stiffness properties) of a scaffold should be tailored for a precise application and have similar strength to the native tissue for which it was designed to support. Figure 11 shows the modulus values of various types of cells and tissues in living organisms [112].

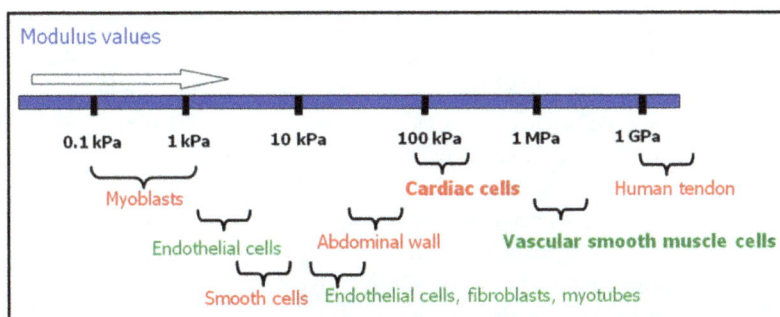

Fig. 11. Range of modulus values observed for different types of cells and tissues in living organisms.

6.2.3 Scaffold architecture; pore size and curvature

Interconnective pore structures with acceptable pore sizes are necessary for scaffolds. The passage of fundamental elements such as nutrients, proteins, gasses, cells, and products within the hydrogel scaffolds is vital [36]. In fact, the pore sizes can be classified according to their diameter size either macro (> 100 μm) or micro (< 100 μm).

To help the bone regeneration, scientists used scaffolds ranging 150-300 μm and 500–710 μm in diameter [109,113-117]. On the other hand, size of pores around 60-100 μm is suited for macrophage cells [114], whereas, pores size less than 150 μm is suitable for burn treatment procedures to induce regeneration of skin [115]. Other than that, reference size of pores to fabricate scaffold for various treatments are also been indicated based on the general cell size including osteoblast (200-400 μm [118]), hepatocyte (82 μm [119]), neovascularization (5 μm [120]), fibrovascular tissue (> 500 μm [121]), and adipocyte (100 μm [122]). Another important characteristic is the curvature of the pore; it generated the best tension and compression on the cells' mechano-receptors, which permitting them to migrate through apertures of such size [109,116].

6.2.4 Biodegradability

Biodegradability is considered among the most critical properties for scaffolds for tissue engineering applications. It is important that the degradability of the scaffold is such that it occurs concurrently with the regeneration of the new tissue. The scaffold must hold its structural integrity for cells homing but will eventually degrade to allow the adhered cells to proliferate, grow and differentiate to form new tissues or organs. A different biodegradation rate is required for different tissue engineering applications.

The rates of degradation mostly depend on the polymer composition of the scaffolds. For polyesters (PLA, PGA, PLGA, PCL), the degradation is driven either by abiotic (chemical hydrolysis and thermal degradation) or biotic (enzymatic and microbial) processes [123,124]. The prime mechanism of biodegradation of PLA-PGA materials involves the nonspecific hydrolytic scission, simple hydrolysis that cleaves the polymer chains at their ester linkages. For PLA, hydrolytic scission disrupts the polymer down to its monomeric form which is then subsequently removed from the body. In addition, the biodegradation of PGA occurs in two approaches, either by hydrolysis or nonspecific esterase and carboxypeptidase activity. The residue which is glycolic acid monomer is either eliminated in the urine or enters the tricarboxylic acid cycle [124].

PEG is well known for its biocompatibility and high hydrophilicity, which has resulted in its extensive utilization as a scaffold in tissue engineering areas. However, PEG is a hydrolytically non-degradable polymer and experiences limited metabolism in the body. Therefore, the entire polymer chain can be removed through the kidneys (< 30 kDa) and liver (> 30 kDa). For that reason, PEG with molecular weights < 50 kDa can be considered for use in tissue engineering applications as optimum removal from the body can be achieved [125]. Copolymerization or cross-linking of PEG with various biodegradable polymers such as PLA, PLGA or PCL will increase the degradation of adjacent polymer chains and therefore, will enhance the degradation rate of the scaffolds.

7.   Methods of improving hydrogels mechanical properties

One of the most widely used methods for reinforcing a hydrogel is through the application of an interpenetrating polymer network (IPN). Various studies have reported that IPNs provide improvements in mechanical properties of hydrogels [126-128]. An IPN is referred as combination of two or more cross-linked polymers. Where one of them is polymerized or cross-linked in the instant existence of the other(s) [129]. IPNs have been broadly studied to combine the properties of a varied range of materials, including plastics, elastomers and hydrogels. Moreover, IPNs lead to an enhancement of each component of polymer properties, resulting in a positive synergistic effect [129,130]. Figure 12 shows the two common classes of IPN; (a) semi IPN, composed of two polymers (one cross-linked and one linear), and (b) full-IPN, composed of two cross-linked polymers.

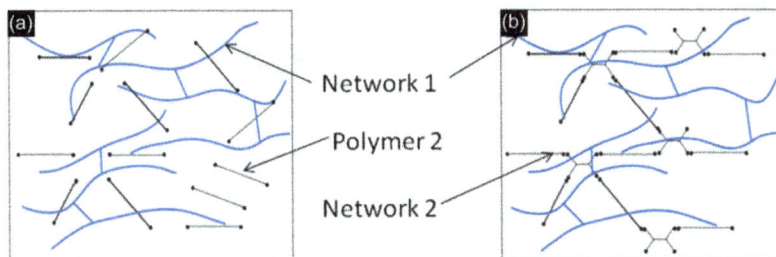

Fig. 12. Schematic of (a) semi and (b) full-IPNs.

However, there are several limitations of IPN networks, especially for semi-IPN hydrogel networks. For example, PHEMA semi-IPNs synthesised in the existence of hydrophilic polymers (i.e: PEG and poly(vinyl pyrrolidone)) [131-133] displayed improved swelling properties, but when the linkages were positioned in water, subsequently the hydrophilic linear polymers could diffuse from the gel matrix. As a result, unwanted volume changes; the introduction of undesirable substances into the surrounding media and reduced the mechanical properties of the hydrogels. Another limitation is IPNs are often formed from non-degradable components, thus they are not applicable for cell culture. Double networks (DNs), a subset of IPNs, were introduced by Gong and co-workers and showed improved mechanical properties [127,134]. The DNs are made up of two hydrophilic networks and are prepared by following approach. A highly cross-linked rigid gel is produced, which is then followed by the production of a loosely cross-linked gel [134].

Most recent approaches that have been employed to enhance the mechanical strength of hydrogels include the introduction of sliding cross-linking agents and nanoparticles (clay-filled composites) [134]. The sliding cross-linking network was first discovered by Okumura and co-workers by chemically cross-linking two or more cyclodextrins, where each threaded on different PEG chains end capped with bulky groups (i.e.: andamantan) [134,135]. This results in the production of a sliding double ring cross-linking agent. Figure 13 shows the topology of this type of cross-linked hydrogel, which contains free moving (sliding mode) chains. Outstanding mechanical properties were observed with high degrees of swelling and a high stretching ratio. A 'pulley effect' causes the sliding double-ring cross-links to equalize the tension along the polymer chains [134,135]. The α-Cyclodextrin moieties threaded on PEG chains are cross-linked by trichlorotriazine producing double and triple ring cross-linkers. This free moving mechanism of the cross-linker is therefore responsible for the improved mechanical properties of the resulting hydrogels.

Materials Research Forum LLC
https://doi.org/10.21741/9781644901892-3

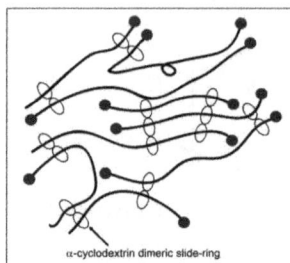

α-cyclodextrin dimeric slide-ring

Fig. 13. Sliding hydrogel structure, consisting of α-Cyclodextrin moieties attached on PEG chains [136].

Haraguchi and co-workers designed novel nanocomposite (clay-filled) hydrogels based on N-isopropylacrylamide (NIPAAm) using hectorite ([$Mg_{5.24}Li_{0.66}Si_8O_{20}(OH)_4$]$Na_{0.66}$) as a multifunctional cross-linker [134]. The architecture of the nanocomposite hydrogel is illustrated in Figure 14. It was determined that the exfoliated clay platelets were uniformly dispersed throughout the hydrogel, and poly(N-isopropylacrylamide) (PNIPAAm) chains were conjugated on the clay platelet surface by one or both ends. The mechanical strength of this organic-inorganic hydrogels was improved due to the flexibility of the PNIPAAm chains holding the platelets together.

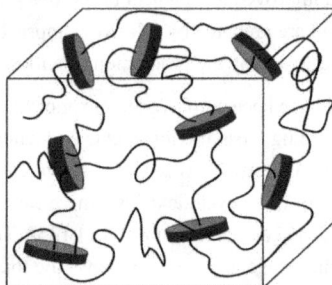

Fig. 14. A schematic structure of the nanocomposite (clay-filled) hydrogel by Haraguchi and co-workers

The approaches discussed offer alternative methods for improving the hydrogels mechanical strength and have been implemented in numerous biomedical areas (i.e: medical surgery, pharmaceutical, and drug delivery fields). Though, it is also vital that the modification made to

Applications of Polymers in Surgery   Materials Research Forum LLC
Materials Research Foundations **123** (2022) 53-91   https://doi.org/10.21741/9781644901892-3

enhance the mechanical properties do not negatively impact on the biocompatibility or other functions of the hydrogel.

8. Tissue-materials responses

The need for designing biomaterials that could be used to replace damaged tissue or organs are desirable as a result of the limited tissue or organ donors. The biomaterials to be utilized within the human body (implanted into living organism) generally undergo tissue responses, which involve various fundamental aspects including wound or damage tissue, inflammatory and wound curing responses, rejection reactions and fibrous encapsulation of the biomaterials, medical devices or prosthesis [137]. Figure 15 shows the immediate responses of biomaterials or medical devices, where they immediately obtain a deposition of host proteins preceding on the biomaterial surfaces. These protein absorptions are reliant on the surface of the biomaterial, which influence the homing and survival of cells. Where macrophages, monocytes, and FBGCs on protein-coated surfaces are involved. The sequence of events for inflammation and the rejection response allows the movement of macrophages or monocytes to the implant spot. The migration of these cells is guided in response to chemokines and other chemo-attractants [137]. Macrophages then adhere to the biomaterials without specifically recognizing it and disperse on its surface as they attempt to phagocytose it [138]. For biomaterials with sizes above the dimension of single macrophage engulfment for particle sizes ranging 10-100 μm in diameter, multinucleated giant cells (MnGCs) or foreign body giant cells (FBGCs) will be developed [139]. The MnGCs and FBGCs will then give out signals demonstrating a large foreign body needs to be confined. Further, the arrival of fibroblasts occurs following the synthesis of collagen capsule, which is mostly directed by the signals from the macrophages [138]. The degradation of the biomaterials are also influenced by the occurrence of FBGCs where the rate of material degradation is higher with the existence of the giant cells [140,141].

Inflammation is the response of vascularised biological tissue to local wound or damage, which involves containing, neutralizing, diluting or walling off the injury agent or process. Overall, the wound healing and the new tissue regeneration processes may be dependent on several factors including shape, size, physical, and chemical characteristics of the biomaterial as well as the size of the artificial device implanted [142]. Generally, biocompatibility is classified into two main classes; acute and chronic inflammatory responses. The main event seen is the fibrous capsule formation over various timepoints following the implantation [142].

IMPLANTATION:

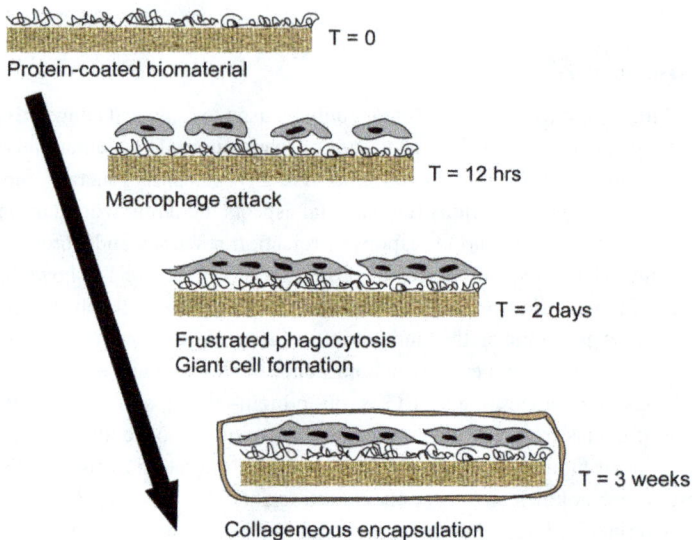

Fig. 15. The time course and generalized reactions that comprise the foreign body reaction.

Macrophages are one of the cell types which produce numerous cytotoxic and microbicidal materials (i.e: superperoxide, nitric oxide and hydrogen peroxide) to kill invading microorganisms [139]. They also secrete groups of compounds which could trigger the rise of body temperature, tissue injury and other inflammatory responses (i.e. clotting factors, cytokines and neutral proteinase). Furthermore, the regulation on innate and the adaption of immune responses are also caused by this macrophage [139]. The tissue restoration and regeneration are negatively impacted by the occurrence of macrophages. Thus, it is vital to point out, besides of their role in the inflammation and immune reaction, macrophages also act a significant function in tissue regeneration. Leibovich and co-workers and Martin have demonstrated that in wound healing processes, the healing is severely impaired if the macrophages are not allowed to invade the damage site [143,144]. However, the persistent occurrence of macrophages, which then fuse into FBGCs would lead to chronic inflammation [142]. In summary, the polymer biomaterials selection for scaffold development or prosthetic device is important to prevent the rejection after the implantation. The biocompatible, non-toxic, adequate mechanical properties with interconnecting pores morphology of the polymer scaffolds will be the succeeding parameters for tissue engineering applications.

9. Summary

Herein, we have shown that there are various potential polymer biomaterials that can be explored for different clinical applications for tissue engineering fields. Broad range of properties offered by the polymer biomaterials leading to numerous developments of scaffolds from flexible to rigid applications. The conventional fabrication techniques have disadvantages such as limited size, limited interconnected porous structure and difficult to fabricate a complex structure of the 3D scaffolds. Thus, the additive manufacturing technology was presented to give insight on the current emerging fabrication technique. The 3D printing technology enables fabrication of 3D complex shaping with specific interconnected pores morphology, design at low expense and faster fabrication rate. This technology too, prevents the wastage of materials used. From the knowledge on the common reaction between biomaterials and cells, consideration for the materials selections is very crucial to ensure the success of the developed polymer scaffolds. Therefore, the combination of suitable polymer biomaterials selection and reliable 3D scaffold fabrication techniques are important to develop an effective polymer scaffold as an implant for tissue engineering applications.

Acknowledgement

The authors would like to extend their acknowledgement to Universiti Sains Malaysia for funding the research under the Research University Grant (Individual) – 1001/PBAHAN/8014133.

References

[1] The Ontario's Organ and Tissue Donation Agency, Toronto, Canada: Annual Report 2009. URL http://www.giftonline.on.ca/page.cfm?id=93C7F131-0C19-48D7-BBBCD444069B220A (accessed 20 September 2021).

[2] Health Resources and Services Administration (HRSA): organdonor.gov 2020. URL https://www.organdonor.gov/learn/organ-donation-statistics/detailed-description#fig1 (accessed 20 September 2021).

[3] Langer, R.; Vacanti, J. P. Tissue Eng. 260 (1993) 920-926. https://doi.org/10.1126/science.8493529

[4] Langer, R. Tissue Eng. 13(1) (2007) 1-2. https://doi.org/10.1089/ten.2006.0219

[5] Ma, P. X. Mater. Today 7(5) (2004) 30-40. https://doi.org/10.1016/S1369-7021(04)00233-0

[6] Prime, E. L.; Hamid, Z. A.; Cooper-White, J. J.; Qiao, G. G. Biomacromolecules 8 (2007) 2416-2421. https://doi.org/10.1021/bm0702962

[7] Wacker, B. K.; Alford, S. K.; Scott, E. A.; Thakur, M. D.; Longmore, G. D.; Elbert, D. L. Biophys. J. 94(1) (2008) 273-285. https://doi.org/10.1529/biophysj.107.109074

[8] Shin, Y. M.; Kim, K. -S.; Lim, Y. M.; Nho, Y. C.; Shin, H. Biomacromolecules 9(7) (2008) 1772-1781. https://doi.org/10.1021/bm701410g

[9] Dikovski, D.; Bianco-Peled, H.; Seliktar, D. Biomaterials 27(8) (2006) 1496-1506. https://doi.org/10.1016/j.biomaterials.2005.09.038

[10] Nikolova, M. P.; Chalavi, M. S. Bioactive Materials 4 (2019) 271-292. https://doi.org/10.1016/j.bioactmat.2019.10.005

[11] Vats, A.; Tolley, N. S.; Polak, J. M.; Gough, J. E. Clinical of Otolaryngol 28(3) (2003) 165-172. https://doi.org/10.1046/j.1365-2273.2003.00686.x

[12] Hutcmacher, D. W. Biomaterials 21(24) (2000) 2529-2543. https://doi.org/10.1016/S0142-9612(00)00121-6

[13] Suh, J. K.; Matthew, H. W. Biomaterials 21(24) (2000) 2589-2598. https://doi.org/10.1016/S0142-9612(00)00126-5

[14] Liao, E.; Yaszemski, M.; Krebsbach, P.; Hollister, S. Tissue Eng. 13(3) (2007) 537-550. https://doi.org/10.1089/ten.2006.0117

[15] Wang, Y.; Kim, U. -J.; Blasioli, D. J.; Kim, H. -J.; Kaplan, D. L. Biomaterials 26(34) (2005) 7082-7094. https://doi.org/10.1016/j.biomaterials.2005.05.022

[16] Pattison, M.; Webster, T. J.; Leslie, J.; Kaefer, M.; Haberstroh, K. M. Macromol. Biosci. 7(5) (2007) 690-700. https://doi.org/10.1002/mabi.200600297

[17] Seal, B. L.; Otero, T. C.; Panitch, A. Mater. Sci. Eng. R. 34(4-5) (2001) 147-230. https://doi.org/10.1016/S0927-796X(01)00035-3

[18] Chan, B. P.; Leong, K. W. European Spine Journal 4 (2008) 467-479. https://doi.org/10.1007/s00586-008-0745-3

[19] Dhandayuthapani, B.; Yoshida, Y.; Maekawa, T.; Kumar, D. S. Int. J. Polym. Sci. (2011). https://doi.org/10.1155/2011/290602

[20] Yukna, R. A.; Yukna, C. N. J. Clin. Periodontol. 25(12) (1998) 1036-1040.
https://doi.org/10.1111/j.1600-051X.1998.tb02410.x

[21] Yukna R. A. J. Periodontol. (1994) 177-185.
https://doi.org/10.1902/jop.1994.65.2.177

[22] Petite H.; Viateau, V.; Bensaid, W.; Meunier, A.; de Pollak, C.; Bourguignon, M.;
Oudina, K.; Sedel, L.; Guillemin, G. Nature Biotechnol. 18(9) (2000) 959-963.
https://doi.org/10.1038/79449

[23] Guillemin, G.; Patat, J. L.; Fournie, J.; Chetail, M. J. Biomed. Mat. Res. 21(5)
(1987) 557-567. https://doi.org/10.1002/jbm.820210503

[24] Piecuch, J. F.; Goldberg, A. J.; Shastry, C. V.; Chrzasnowski, R. B. J. Biomed. Mat.
Res. 18(1) (1984) 39-45. https://doi.org/10.1002/jbm.820180106

[25] Ma, Z.; Gao, H.; Gong, Y.; Shen, J. J. Biomed. Mater. Res. Part B: Appl. Biomater.
67B(1) (2003) 610-617. https://doi.org/10.1002/jbm.b.10049

[26] Holland, T. A.; Mikos, A. G. Adv. Biochem. Eng. Biotechnol. 102 (2006) 161-185.

[27] Sawtell, R. M.; Kayser, M. V.; Downes, S. Cell Materials (1995) 63-71.

[28] Nehrer, S.; Breinan, H. A.; Ramappa, A.; Hsu, H. P.; Minas, T.; Shortkroff, S.;
Sledge, C. B.; Yannas, I. V.; Spector, M. Biomaterials 19(24) (1998) 2313-2328.

[29] Risbud, M.; Ringe, J.; Bhonde, R.; Sittinger, M. Cell Transplantation 10(8) (2001)
755-763. https://doi.org/10.3727/000000001783986224

[30] Brown, A. N.; Kim, B. -S.; Alsberg, E.; M. S. E.; Mooney, D. J. Tissue Eng. 6(4)
(2000) 297-305. https://doi.org/10.1089/107632700418029

[31] Huang, X.; Zhang, Y.; Donahue, H. J.; Lowe, T. L. Tissue Eng. 13(11) (2007) 2645-
2651. https://doi.org/10.1089/ten.2007.0084

[32] Zhang, L.; Hu, J.; Athanasiou, A. Crit. Rev. Biomed. Eng. 37(1-2) (2009) 1-57.
https://doi.org/10.1615/CritRevBiomedEng.v37.i1-2.10

[33] Richbourg, N. R.; Peppas, N. A.; Sikavitsas, V. I. J. Tissue Eng. Regen. Med. 13(8)
(2019) 1275-1293. https://doi.org/10.1002/term.2859

[34] Dickinson, L. E.; Gerecht, S.; Front. Physiol. 7 (2016) 341.
https://doi.org/10.3389/fpls.2016.00341

[35] Tottoli, E. M.; Dorati, R.; Genta, I.; Chiesa, E.; Pisani, S.; Conti, B. Pharmaceutics
12(8) (2020) 735. https://doi.org/10.3390/pharmaceutics12080735

[36] Drury, J. L.; Mooney, D. J. Biomaterials 24(24) (2003) 4337-4351.
https://doi.org/10.1016/S0142-9612(03)00340-5

[37] Xu, L.; Liu, L.; Liu, F.; Cai, H.; Zhang, W. Polym. Chem. 6(15) (2015) 2945-2954.
https://doi.org/10.1039/C5PY00039D

[38] Loebsack, A.; Greene, K.; Wyatt, S.; Culberson, C.; Austin, C.; Beiler, R.; Roland,
W.; Eiselt, P.; Rowley, J.; Burg, K.; Mooney, D.; Holder, W.; Halberstadt, C. J.
Biomed. Mater. Res. 57(4) (2001) 575-581. https://doi.org/10.1002/1097-
4636(20011215)57:4<575::AID-JBM1204>3.0.CO;2-9

[39] VandeVord, P. J.; Matthew, H. W. T.; DeSilve, S. P.; Mayton, L.; Wu, B.; Wooley,
P. H. J. Biomed. Mater. Res. 59(3) (2002) 585-590. https://doi.org/10.1002/jbm.1270

[40] Ono, K.; Saito, Y.; Yura, H.; Ishikawa, K; Kurita, A.; Akaike, T.; Ishihara, M. J.
Biomed. Mater. Res. 49(2) (2000) 289-295. https://doi.org/10.1002/(SICI)1097-
4636(200002)49:2<289::AID-JBM18>3.0.CO;2-M

[41] Zhao, X.; Kato, K.; Fukumoto, Y.; Nakamae, K. Int. J. Adhes. and Adhes. 21(3)
(2001) 227-232. https://doi.org/10.1016/S0143-7496(01)00003-3

[42] Yao, Rui.; Zhang, Renji.; Yan, Y.; Wang, X.; Journal of Bioactive and Compatible
Polymers 24(1) (2009) 48-62. https://doi.org/10.1177/0883911509103329

[43] Patrick, Jr. C. W. Semin. Surg. Oncol. 19(3) (2000) 302-311.
https://doi.org/10.1002/1098-2388(200010/11)19:3<302::AID-SSU12>3.0.CO;2-S

[44] Gomillion, C. T.; Burg, K. J. L. Biomaterials 27(36) (2006) 6052-6063.
https://doi.org/10.1016/j.biomaterials.2006.07.033

[45] Kononas, T. C.; Bucky, L. P.; Hurley, C. May, J. W. Plast. Reconstr. Surg., 91(5)
(1993) 763-768. https://doi.org/10.1097/00006534-199304001-00001

[46] Bartynski, J.; Marion, M. S.; Wang, T. D. Otalaryngol Head Neck Surg. 102(4)
(1990) 314-321. https://doi.org/10.1177/019459989010200402

[47] Sommer, B.; Sattler, G. Dermatol. Surg. 26(12) (2000) 1159-1166.
https://doi.org/10.1046/j.1524-4725.2000.00278.x

[48] Kimura, Y.; Ozeki, M.; Inamoto, T.; Tabata, Y. Biomaterials 24(14) (2003) 2513-
2521. https://doi.org/10.1016/S0142-9612(03)00049-8

[49] Vashi, A. V.; Keramidaris, E.; Abberton, K. M.; Morrison, W. A.; Wilson, J. L.;
O'Connor, A. J.; Cooper-White, J. J.; Thompson, E. W. Biomaterials 29(5) (2008)
573-579. https://doi.org/10.1016/j.biomaterials.2007.10.017

[50] Flynn, L.; Woodhouse, K. A. Organogenesis 4(4) (2008) 228-235. https://doi.org/10.4161/org.4.4.7082

[51] Green, H.; Kehinde, O. Cell 7(1) (1976) 105-113. https://doi.org/10.1016/0092-8674(76)90260-9

[52] Green, H.; Meuth, M. Cell 3(2) (1974) 127-133. https://doi.org/10.1016/0092-8674(74)90116-0

[53] Kuri-Harcuch, W.; Green, H. Proc. Natl. Acad. Sci. USA 75(12) (1978) 6107-6109. https://doi.org/10.1073/pnas.75.12.6107

[54] Xu, F.; Gomillion, C. T.; Maxson, S.; Burg, K. J. L. J. Tissue Eng. Regen. Med. 3 (2009) 338-347. https://doi.org/10.1002/term.158

[55] Conrad, C.; Huss, R. J. Surg. Res. 124(2) (2005) 201-208. https://doi.org/10.1016/j.jss.2004.09.015

[56] Pelled, G. G. T.; Aslan, H.; Gazit, Z.; Gazit, D. Curr. Pharm. Des. 8(21) (2002) 1917-1928. https://doi.org/10.2174/1381612023393666

[57] Alhadlaq, A.; Tang, M.; Mao, J. J. Tissue Eng. 11(3-4) (2005) 556-566. https://doi.org/10.1089/ten.2005.11.556

[58] Keskar, V. B. S.; Marion, N. W.; Mao, J. J.; Gemeinhart, R. A. Tissue Eng. 15(7) (2009) 1695-1707. https://doi.org/10.1089/ten.tea.2008.0238

[59] Elisseeff, J.; Anseth, K.; Sims, D.; Mcintosh, W.; Randolph, M.; Yaremchuk, M.; Langer, R. Plast. Reconstr. Surg. 104(4) (1999) 1014-1022. https://doi.org/10.1097/00006534-199909020-00018

[60] Burdick, J. A.; Anseth, K. S. Biomaterials 23(22) (2002) 4315-4323. https://doi.org/10.1016/S0142-9612(02)00176-X

[61] Pradhan, S.; Hassani, I.; Seeto, W. J.; Lipke, E. A. J. Biomed. Mater. Res. A. 105(1) (2017) 236-252. https://doi.org/10.1002/jbm.a.35899

[62] Abberton, K. M.; Bortolotto, S. K.; Woods, A. A.; Findlay, M.; Morrison, W. A.; Thompson, E. W.; Messina, A. Cells Tissues Organs 188(4) (2008) 347-358. https://doi.org/10.1159/000121575

[63] Kawaguchi, N.; Toriyama, K.; Nicodemou-Lena, E.; Inou, K.; Torii, S.; Kitagawa, Y. Proc. Natl. Acad. Sci. USA 95(3) (1998) 1062-1066. https://doi.org/10.1073/pnas.95.3.1062

[64] Cronin, K. J.; Messina, A.; Knight, K. R.; et al. Plast. Reconstr. Surg. 113(1) (2004) 260-269. https://doi.org/10.1097/01.PRS.0000095942.71618.9D

[65] Chenite, A.; Chaput, C.; Wang, D.; Combes, C.; Buschman, M. D.; Hoemann, C. D.; Leroux, J. C.; Atkinson, B. L.; Binette, F.; Selmani, A. Biomaterials 21(21) (2000) 2155-2161. https://doi.org/10.1016/S0142-9612(00)00116-2

[66] Lee, K. Y.; Mooney, D. J. Chemical Reviews 101(7) (2001) 1869-1880. https://doi.org/10.1021/cr000108x

[67] Buckley, C. T.; O'Kelly, K. U. Topics in Biomechanical Engineering. Trinity Centre for Bioengineering (2004) p147-166.

[68] Balakrishnan, -B.; Jayakrishnan, A. Biomaterials 26(18) (2005) 3941-3951. https://doi.org/10.1016/j.biomaterials.2004.10.005

[69] Yoo, H. S.; Lee, E. A.; Yoon, J. J.; Park, T. G. Biomaterials 26(14) (2005) 1925-1933. https://doi.org/10.1016/j.biomaterials.2004.06.021

[70] Eiselt, P.; Lee, K. Y.; Mooney, D. J. Macromolecules 32(17) (1999) 5561-5566. https://doi.org/10.1021/ma990514m

[71] Lutoft, M. P.; Weber, F. E.; Schmoekel, H. G.; Schense, J. C.; Kohler, T.; Muller, R.; Hubbell, J. A. Nat. Biotechnol. 21(5) (2003) 513-518. https://doi.org/10.1038/nbt818

[72] Kim, S.; Healy, K. E. Biomacromolecules 4(5) (2003) 1214-1223. https://doi.org/10.1021/bm0340467

[73] Liu, C.; Xia, Z.; Chernuszka, J. T. Trans IChemE, Part A, Chemical Engineering Resaerch and Design 85(A7) (2007) 1051-1064. https://doi.org/10.1205/cherd06196

[74] Wong, W. H.; Mooney, D. J. Synthetic Biodegradable Polymer Scaffolds. Birkhauser Boston (1997) p 51-82. https://doi.org/10.1007/978-1-4612-4154-6_4

[75] Jagur-Grodzinski, J. Polym. Adv. Technol. 17(6) (2006) 395-418. https://doi.org/10.1002/pat.729

[76] Oh, S. H.; Kang, S. G.; Kim, E. S.; Cho, S. H.; Lee, J. H. Biomaterials 24(22) (2003) 4011-4021. https://doi.org/10.1016/S0142-9612(03)00284-9

[77] Pradny, M.; Lesny, P.; Fiala, J.; Vacik, J.; Slouf, M.; Michalek, J.; Sykova, E. Collect. Czech. Chem. Commun. 68(4) (2003) 812-822. https://doi.org/10.1135/cccc20030812

[78] Langer, R. J. Controlled Release 62 (1999) 7-11. https://doi.org/10.1016/S0168-3659(99)00057-7

[79] Mikos, A. G.; Thorsen, A. J.; Cherwonka, L. A.; Bao, Y.; Langer, R.; Winslow, D. N.; Vacanti, J. P. Polymer 35(5) (1994) 1068-1077. https://doi.org/10.1016/0032-3861(94)90953-9

[80] Lu, L.; Mikos, A. G. MRS Bulletin 21(11) (1996) 28-32. https://doi.org/10.1557/S088376940003181X

[81] Gao, J.; Crapo, P. M.; Wang, Y. Tissue Eng. 12(4) (2006) 917-925. https://doi.org/10.1089/ten.2006.12.917

[82] Nam, Y. S.; Park, T. G. J. Biomed. Mater. Res. 47(1) (1999) 8-17. https://doi.org/10.1002/(SICI)1097-4636(199910)47:1<8::AID-JBM2>3.0.CO;2-L

[83] Mikos, A. G.; Bao, Y.; Cima, L. G.; Ingeber, D. E.; Vacanti, J. P.; Langer, R. J. Biomed. Mater. Res. 27(2) (1993) 183-189. https://doi.org/10.1002/jbm.820270207

[84] Hou, Q.; Grijpma, D. W.; Feijin, J. Biomaterials 24(11) (2003) 1937-1947. https://doi.org/10.1016/S0142-9612(02)00562-8

[85] Nam, Y. S.; Park, T. G. Biomaterials 20(19) (1999) 1783-1790. https://doi.org/10.1016/S0142-9612(99)00073-3

[86] Dhariawala, B.; Hunt, E.; Boland, T. Tissue Eng. 10(9-10) (2004) 1316-1322. https://doi.org/10.1089/ten.2004.10.1316

[87] Gomes, M. E.; Ribeiro, A. S.; Malafaya, P. B.; Reis, R. L.; Cunha, A. M. Biomaterials 22(9) (2001) 883-889. https://doi.org/10.1016/S0142-9612(00)00211-8

[88] Freed, L. E.; Marquis, J. C.; Nohria, A.; Emmanual, J.; Mikos, A. G.; Langer, R. J. Biomed. Mater. Res. 27(1) (1993) 11-23. https://doi.org/10.1002/jbm.820270104

[89] Nam, Y. S.; Yoon, J. H.; Park, T. G. J. Biomed. Mater. Res. Part B. 53 (2000) 1-7. https://doi.org/10.1002/(SICI)1097-4636(2000)53:1<1::AID-JBM1>3.0.CO;2-R

[90] Draghi, L.; Resta, S.; Pirozzolo, M. G.; Tanzi, M. C. J. Mater. Sci.: Mater. in Med. 16 (2005) 1093-1097. https://doi.org/10.1007/s10856-005-4711-x

[91] Lo, H.; Ponticiello, M. S.; Leong, K. W. Tissue Eng. 1(1) (1995) 15-28. https://doi.org/10.1089/ten.1995.1.15

[92] Schugens, C.; Maquet, V.; Grandfils, C.; Jerome, R.; Teyssie, P. J. Biomed. Mater. Res. 30(4) (1996) 449-461. https://doi.org/10.1002/(SICI)1097-4636(199604)30:4<449::AID-JBM3>3.0.CO;2-P

[93] Mikos, A. G.; Temenoff, J. S. J. Biotechnol. 3(2) (2000) 114-119. https://doi.org/10.2225/vol3-issue2-fulltext-5

[94] Petrovic, V.; Vicente, J.; Ramn. J.; Portols, L. Biomedicine (2012).

[95] Rastogi, P.; Kandasubramanian, B. Chem. Eng. J. 366 (2019) 264-304. https://doi.org/10.1016/j.cej.2019.02.085

[96] Butt, J.; Bhaskar, R. J. Manuf. Mater. Process 4(2) (2020) 1-20.

[97] Gómez-Ciriza, G.; Hussain, T.; Gómez-Cía, T.; Valverde, I. Interv. Cardiol. 7(4) (2015) 343-350. https://doi.org/10.2217/ica.15.25

[98] Negi, S.; Dhiman, S.; Sharma, R. K. Rapid Prototyp. J. 20(3) (2014) 256-267. https://doi.org/10.1108/RPJ-07-2012-0065

[99] Ramola, M.; Yadav, V.; Jain, R. J. Manuf. Technol. Manag. 30(1) (2019) 48-69. https://doi.org/10.1108/JMTM-03-2018-0094

[100] Ghilan, A.; Chiriac, A. P.; Nita, L. E.; Rusu, A. G.; Neamtu, I.; Chiriac, V. M. J. Polym. Environ. 28(5) (2020) 1345-1367. https://doi.org/10.1007/s10924-020-01722-x

[101] Ursan, I.; Chiu, L.; Pierce, A. J. Am. Pharm. Assoc. 53(2) (2013) 135-44. https://doi.org/10.1331/JAPhA.2013.12217

[102] Jamróz, W.; Szafraniec, J.; Kurek, M.; Jachowicz, R. Pharm. Res. 35(9) (2018). https://doi.org/10.1007/s11095-018-2454-x

[103] Mertz, L. IEEE Pulse 4(6) (2013) 15-21. https://doi.org/10.1109/MPUL.2013.2279616

[104] Gross, B. C.; Erkal, J. L.; Lockwood, S. Y.; Chen, C.; Spence, D. M. Anal. Chem. 86(7) (2014) 3240-53. https://doi.org/10.1021/ac403397r

[105] Hoy, M. B. Med. Ref. Serv. Q. 32(1) (2013) 94-9.

[106] Cevik, U.; Kam, M. Journal of Nanomaterials (2020).

[107] Banks, B. J. IEEE Pulse 4(6) (2013) 22-6. https://doi.org/10.1109/MPUL.2013.2279617

[108] Schubert, C.; van Langeveld, M. C.; Danoso, L. A. Br. J. Ophthalmol. 98(2) (2014) 159-61. https://doi.org/10.1136/bjophthalmol-2013-304446

[109] Abdul Hamid, Z. A.; Ismail, H.; Ahmad, Z. Materials Science Forum 819 (2015) 361-366. https://doi.org/10.4028/www.scientific.net/MSF.819.361

[110] LaPorte, R. J., Hydrophilic Polymer Coatings for Medical Devices. Technomic Publishing Company, Lancaster, PA (1997) p 20.

[111] Shu, X. Z.; Liu, Y.; Palumbo, F. S.; Luo, Y.; Prestwich, G. D. Biomaterials 25(7-8) (2004) 1339-1348. https://doi.org/10.1016/j.biomaterials.2003.08.014

Materials Research Forum LLC
https://doi.org/10.21741/9781644901892-3

[112] Moschou, E. A.; Chopra, N.; Khatwani, S. L.; Ehrick, J. D.; Deo, S. K.; Bachas, L. G.; Daunert, S. Mater. Res. Soc. Symp. Proc. (2007) 952. https://doi.org/10.1557/PROC-0952-F05-02

[113] Ishaug-Riley, S. L.; Crane-Kruger, G. M.; Yaszemski, M. J.; Mikos, A. G. Biomaterials 19(15) (1998) 1405-1412. https://doi.org/10.1016/S0142-9612(98)00021-0

[114] Mikos, A. G.; Sarakinos, G.; Lyman, M. D.; Ingber, D. E.; Vacanti, J. P.; Langer, R. Biotechnol. Bioeng. 42(6) (1993) 716-723. https://doi.org/10.1002/bit.260420606

[115] O'Brien, F. J.; Harley, B. A.; Yannas, I. V.; Gibson, L. Biomaterials 25(6) (2004) 1077-1086.

[116] Boyan, B. D.; Hummert, T. W.; Dean, D. D.; Scwartz, Z. Biomaterials 17(2) (1996) 137-146. https://doi.org/10.1016/0142-9612(96)85758-9

[117] O'Brien, F. J.; Harley, B. A.; Yannas, I. V.; Gibson, L. J. Biomaterials 26(4) (2005) 433-441. https://doi.org/10.1016/j.biomaterials.2004.02.052

[118] Chang, B. S.; Lee, C. K.; Hong, K. S.; Youn, H. J.; Ryu, H. S.; Chung, S. S.; Park, K. W. Biomaterials 21(12) (2000) 1291-1298. https://doi.org/10.1016/S0142-9612(00)00030-2

[119] Ranucci, C. S.; Kumar, A.; Batra, S. P.; Moghe, P. V. Biomaterials 21(8) (2000) 783-793. https://doi.org/10.1016/S0142-9612(99)00238-0

[120] Brauker, J. H.; Carr-Brendel, W. E.; Martinson, L. A.; Crudele, J.; Johnston, W. D.; Johnson, R. C. J. Biomed. Mater. Res. 29 (1995) 1517-1524. https://doi.org/10.1002/jbm.820291208

[121] Wake, M. C.; Patrick, C. W.; Mikos A. G. Cell Transplantation 3(4) (1994) 586-597. https://doi.org/10.1177/096368979400300411

[122] Kumerta, M.; Heimburg, D. V.; Schoof, H.; Heschel, I.; Rau, G. Int. J. Artif. Organs 25(1) (2002) 67-73. https://doi.org/10.1177/039139880202500111

[123] Griffith, L. G. Acta Biomater. 48(1) (2000) 263-277. https://doi.org/10.1016/S1359-6454(99)00299-2

[124] Hakkarainen, M. Adv. Polym. Sci. 157 (2002) 113-118. https://doi.org/10.1007/3-540-45734-8_4

[125] Tessmar, J. K.; Gopferich, A. M. Macromolecular Science 7(1) (2007) 23-39. https://doi.org/10.1002/mabi.200600096

[126] Myung, D.; Koh, W.; Ko, J.; Hu, Y.; Carrasco, M.; Noolandi, J.; Ta, C. N.; Frank, C. W. Polymer 48(18) (2007) 5376-5387. https://doi.org/10.1016/j.polymer.2007.06.070

[127] Gong, J. P.; Katsuyama, Y.; Kurokawa, T.; Osada, Y. Adv. Mater. 15(14) (2003) 1155-1158. https://doi.org/10.1002/adma.200304907

[128] Yin, L.; Fei, L.; Tang, C.; Yin, C. Polym. Int. 56(12) (2007) 1563-1571. https://doi.org/10.1002/pi.2306

[129] Gupta, N.; Srivastava, A. K. Polym. Int. 35(2) (1994) 109-118. https://doi.org/10.1002/pi.1994.210350201

[130] Peng, H. T.; Martineau, L.; Shek, P. N. J. Mater. Sci. - Mater. Med. 18(6) (2007) 975-986. https://doi.org/10.1007/s10856-006-0088-8

[131] Bajpai, A. K.; Shrivastava, M. J. J. Macromol. Sci. Part A Pure Appl. Chem. A39(7) (2002) 667. https://doi.org/10.1081/MA-120004511

[132] Caliceti, P.; Veronese, F.; Schiavon, O.; Lora, S.; Carenza, M. Farmaco 47(3) (1992) 275-286.

[133] Hu, H.; Briggs, C. R.; Nguyen, T.; Tran, H.; Rossberg, E. PCT Patent 15521 (1998).

[134] Kopecek, J.; Yang, J. Polym. Int. 56(9) (2007) 1078-1098. https://doi.org/10.1002/pi.2253

[135] Tanaka, Y.; Gong, J. P.; Osada, Y. Prog. Polym. Sci. 30(1) (2005) 1-9. https://doi.org/10.1016/j.progpolymsci.2004.11.003

[136] Sachlos, E.; Reis, N.; Ainsley, C.; Derby, B.; Czernuszka, J. T. Biomaterials 24(8) (2003) 1487-1497. https://doi.org/10.1016/S0142-9612(02)00528-8

[137] Anderson, J. M.; Rodriguez, A.; Chang, D. T. Seminars in Immunology 20(2) (2008) 86-100. https://doi.org/10.1016/j.smim.2007.11.004

[138] Ratner, B. D. J. Controlled Release 78(1-3) (2002) 211-218. https://doi.org/10.1016/S0168-3659(01)00502-8

[139] Xia, Z.; Triffitt, J. T. Biomed. Mater. 1(1) (2006) R1-R9. https://doi.org/10.1088/1748-6041/1/1/R01

[140] Xia, Z.; Ye, H.; Choong, C.; Ferguson, D. J.; Platt, N.; Cui, Z.; Triffitt, J. T. J. Biomed. Mater. Res. 71A(3) (2004) 538-548. https://doi.org/10.1002/jbm.a.30185

Materials Research Forum LLC
https://doi.org/10.21741/9781644901892-3

[141] Zhao, Q.; Topham, N.; Anderson, J. M.; Hiltner, A.; Lodoen, G.; Payet, C. R. J. Biomed. Mater. Res. 25(2) (1991) 177-183. https://doi.org/10.1002/jbm.820250205

[142] Anderson, J. M. Annu. Rev. Mater. Res. 31 (2001) 81-110. https://doi.org/10.1146/annurev.matsci.31.1.81

[143] Leibovich, S. J.; Ross, R. Am. J. Pathol. 78(1) (1975) 71-100.

[144] Martin, P. Science 276(5309) (1997) 75-81. https://doi.org/10.1126/science.276.5309.75

Materials Research Forum LLC
https://doi.org/10.21741/9781644901892-4

Chapter 4

Applications of Polymers in Neurosurgery

Zarsha Naureen[1], Muhammad Zubair[1†], Fiaz Rasul[2], Habibullah Nadeem[1],
Muhammad Hussnain Siddique[1], Muhammad Afzal[1] and Ijaz Rasul[1*]

[1]Department of Bioinformatics and Biotechnology, Government College University, Allam Iqbal
Road, Faisalabad, 38000, Pakistan

[2]BioAtlantis Ltd., Clash Industrial Estate, Tralee, Co-Kerry, Ireland

*ijazrasul@gcuf.edu.pk

† Contributed equally as first co-author

Abstract

The advancement in regenerative medicine has made it possible to offer new strategies and novel materials that have brought a revolution in the areas of drug delivery, tissue engineering, and cosmetic surgeries. The area of tissue engineering has mainly focused to develop such materials that can mimic the natural extracellular matrix (ECM). The material used widely for this purpose as well as for scaffolds for temporary support is known as a polymer. Polymers both natural and synthetic are being used as biomaterial and it has brought about significant changes in the area of neuronal repair. This chapter discusses the potential areas in neurosurgery where the application of polymers is possible.

Keywords

Polymers, Neurosurgery, Peripheral Nerve Injury, Spinal Injury, Dural Repair

Contents

Applications of Polymers in Surgery   Materials Research Forum LLC
Materials Research Foundations **123** (2022) 92-122   https://doi.org/10.21741/9781644901892-4

Applications of Polymers in Surgery   Materials Research Forum LLC
Materials Research Foundations **123** (2022) 92-122   https://doi.org/10.21741/9781644901892-4

1.   Introduction

The past decade has observed advancements in the field of regenerative medicine by making progress in the area of polymer science. That offers novel materials and new strategies which are revolutionizing drug delivery, tissue engineering, and cosmetic surgery. Research for tissue engineering has been mainly focused on developing new materials that can be used as the natural extracellular matrix (ECM) substitute. That also included the scaffolds which may provide three-dimensional support temporarily, acting as guidance and promote the correct formation of the tissue pattern [1]. Such applications use both natural and synthetic polymers as the biomaterial. They have become the natural choice due to many reasons. These are readily available, easily reproducible, biodegradable and easily adapted for many applications due to their versatility [2].

As the understanding regarding the nervous system is increasing along with the availability of biomaterials, the landscape has significantly changed for developing strategies to repair the nervous system. It has now become possible to tailor biomaterials directly for specific applications with the help of new staining and imaging techniques as well as with the development of new novel substances [3]. The ability of the CNS is extremely limited when it comes to regeneration after the occurrence of an injury. This is because the environment present for the nerve repair is severely hostile. However, there is some cell replacement after the injury to CNS but the neuronal circuit does not have the capacity to reinstate [4,5]. The repair strategies for CNS get benefit from biomaterials due to their ability to develop hydrogel matrices and scaffolds. That plays a critical role in supporting the tissues physically. That bridges the neural gap and guides the bioactive materials or cells through the construct with the help of mechanical cues. This creates a possibility of delivery systems of sustained nature responsible for the migration of cells and other necessary molecules for nourishment. It maximizes the outcome of repair [6–9]. In this chapter, we will discuss applications of different polymers in the field of neurosurgery.

2.   Use of polymers in NGCs in the peripheral nerve injury

2.1   Nerve guidance conduits (NGCs)

For the repair and regeneration of nerves, one of the multidimensional, multidisciplinary, and cutting edge platforms for transformative material that helps in the development of scaffolds based on biomaterial, is represented by NGCs. The tissue engineering properties and principles may be applied in order to make that biomaterial which is biocompatible, biodegradable, and mimics the natural mechanics [10]. It is roughly estimated that out of all trauma patients, almost 2.8% patients have peripheral nerve injury (PNI). Based on classification and severity of the injury, it might be extremely disturbing [11]. However,

PNI has a low incidence but it is significantly related to high morbidity rate. The outcomes have been poor historically for peripheral nerve repair especially in the case when there is severe damage or treatment is delayed [12].

In fact a nerve guidance conduit (NGC) is an artificial nerve graft that connects two ends of a defective nerve, i.e., the proximal and distal ends. Further improvement in the effect of repair and the nerve defect length can be achieved when growth factors and/or cells are combined with nerve conduits. That is also called as nerve tissue engineering grafts. Nerve transplantation is substituted by this procedure. The damages and limitations that result from nerve grafting and nerve transfer may be avoided by this procedure. The structural support required for axons is not only provided by the NGCs, various types of nerve factors, and support for other regenerative-environment is also provided by it. Thus, nerve generation is provided by it [13,14].

2.2 Different types of polymers used in the NGCs

Functional recovery is promoted in injured nerves by using a range of biomaterials. A variety of materials ranging from biopolymers to synthetic ones to the blends are used according to their properties such as biocompatibility, mechanical strength, their capability to carry with themselves different proteins and growth factors by encapsulating them and their degradation profile. Biocompatibility can be obtained from biopolymers which are often designed in such a way that they have the same degradation profiles and mechanical properties which are similar to that of regenerating nerve. Moreover, biopolymers are also responsible to carry and present ECM proteins and growth factors to the proximal end of the nerve [15]. Although natural polymers have good biocompatibility but they have poor mechanical properties. They are also difficult to purify as well. NGCs made from synthetic polymers are relatively easy to make. While NGCs based on synthetic polymers have good mechanical properties. Contrary to that their biocompatibility is not good in comparison to natural polymers [16].

2.2.1 Natural polymers used in the NGCs

The cellular histocompatibility and degradation profile of natural polymers are excellent [17,18]. Owing of that reason, they have wide applications in tissue engineering including peripheral nerve regeneration. Nevertheless, their mechanical properties are weak and the risk of disease transmission and antigenicity is obvious. In the manufacturing of NGCs, several natural polymers are used such as alginate [19,20], hyaluronic acid [21], collagen [22], silk fibroin (SF) [23], and gelatin [24].

2.2.1.1 Hyaluronic acid

The essential component that makes up the extracellular matrix is hyaluronic acid (HA). This molecule also has low immunogenicity. The scaffolds made from HA are most suitable for carrying bioactive substances and drugs because of their porous structure [25]. The HA binds the peptides or adhesion molecules thereby increasing the cell adhesion [21].

2.2.1.2 Collagen

The main and important components that make up the peripheral nerves are collagen type I, II, and III [26,27]. The extracellular matrix is also stimulated by collage, structurally and functionally. Both *in vivo* and *in vitro* demonstrations have been shown for the promotion of myelination and regeneration of axons by the NGCs made from different types of collagen. US Food and Drug Administration (USDA) has approved several products made of collagen but these are restricted for patients having nerve defects less than 3cm [28,29].

2.2.1.3 Silk fibroin

The fibrin that is obtained from the silkworm is silk fibroin (SF) which is also used to fabricate many NGCs for peripheral nerves. The gels, films, and sponges are produced using the SF. The methacrylate-grouped silk fibroin is used in a study. by Kim et al. There its properties of free-radical polymerization and photoactivation are used to produce properties of degradation [23].

2.2.1.4 Gelatin

It is denatured collagen that is widely used for scaffolds used in peripheral nerves. Excellent degradability, good cell compatibility and promotion of regeneration of axon are evident from several studies for nerve ducts made from gelatin [24]. NGCs are printed in the 3D by using methacryloyl gelatin (GelMA). It is obtained by substituting the methacryloyl group in gelation by functionalization. The polymerization functions and photocrosslinking in GelMA are because of this substitute group. Additionally, this substitute group also improves the mechanical properties [24,30].

2.2.2 Synthetic polymers

The convenience with which synthetic polymers are produced is among one of their advantages. Additionally, their mechanical properties are also adjustable. However, compared to natural polymers, their biocompatibility is low. The Poly(caprolactone) PCL [31], poly(ethylene glycol) (PEG) [32,33], poly(lactic acid-co-glycolic acid) (PLGA) [34], poly(glycerol sebacate) (PGS) [35], carbon nanotubes (CNTs) [36], and polypyrrole (PPy)

[22] are some examples of synthetic polymers that have been used either alone or combined with natural polymers for printing NGCs via 3D-printing.

2.2.2.1 PCL

The excellent mechanical properties of PCL make it the most used material to prepare NGCs [37–40]. However, this synthetic polymer has a poor degradation profile.

2.2.2.2 Poly(lactic acid-co-glycolic acid) (PLGA)

The degradation profile of PLGA is adjustable compared to the PCL as well as their mechanical properties. *In vitro* studies have been carried out for the combination of chitosan/PLGA and PLGA/Pluronic F127 NGCs for the promotion of regeneration of axon and recovery of motor function [41,42].

2.2.2.3 Poly(ethylene glycol) (PEG)

The biocompatibility of PEG is excellent but it is not clear how it produces its effect on the regeneration of nerves. However, it is observed that during the NGCs preparation, PEG and PEGDA that are photo-cross-linkable may act as cell carriers. It is also estimated that good cellular activity is present in approximately 87% of the embedded cells [43]. Therefore, PEGDA has a wide application in the addictive fabrication of NGCs based on PEG.

2.2.2.4 Poly(glycerol sebacate) (PGS)

This material has absorbable and photo-curable properties due to which it can be used for printing of NGCs through a 3D printer [35]. NGCs prepared from PGS show greater flexibility as well as similar mechanical properties compared to peripheral nerves. Better adhesion and proliferation are promoted by NGCs made from flat-sheet PGS as compared to PLGA. Moreover, that decreases the inflammatory response. However, it lacks the 3D structure that of the extracellular matrix. Excellent mechanical properties as well as customization according to individual requirements is present in NGCs based on PGS due to their photo-curable nature [35].

2.2.2.5 Carbon nanotubes (CNTs)

The CNTs along with conductive polymers mainly form the conductive material in NGCs of the latest generation [22,36]. The electrical signal and its integrity are maintained by them in the nerve pathway. Furthermore, electrical stimulation is synergized by them as well. Nevertheless, their process of preparation is difficult, they are insoluble and their degradation profile is poor as well. This is the reason why they are combined with natural polymers [44]. A summary of applications of natural biopolymers in NGCs has been explained in Table 1.

Table 1: Use of natural biopolymers in fabricating NGCs

Natural Polymers	Use in NGCs
Hyaluronic acid	One of the most suitable molecules to create porous NGCs for carrying drugs and other bioactive molecules [25]
Collagen (I, II, III)	Collagen creates NGCs for axon regeneration and promotion of myleination [28]
Silk fibroin	Fabrication of NGCs for peripheral nerve injuries [23]
Gelatin	NGCs made from gelatin are used for peripheral nerves, regeneration of axons, along with excellent biodegradability [24]

2.3 NGCs in peripheral nerve regeneration

The process of regeneration of the nerve is complicated because in which there is a significant role of interaction between cell and cell, and cell and extracellular matrix. For maintaining acceptance NGCs by the immune system, their structural integrity, remodeling of tissues, their ability to be implanted over a joint, and their biological performance, NGCs are ideally made to be biocompatible, flexible, biodegradable, properties similar to ECM as well as mechanical stability [45]. Up till now, studies have been carried out to develop strategies for developing ideal NGCs such as the synthesis of biopolymers, their modification, physical stimuli, and their use, bioactive materials/cells incorporation, and manipulating target cells genetically or a combination of them [29]. NGCs are used for repairing long-distanced defects in peripheral nerves where there is a need to extend the axon across the long gap of the nerve and then connecting it to the segment of the distal nerve. The exchange of gas and nutrients is promoted when NGCs are sufficiently permeable. It allows in-growth of capillary, and fibrosis of the repair site is prevented by it through a barrier maintained between the fibroblasts and other cells [42]. (See figure 1).

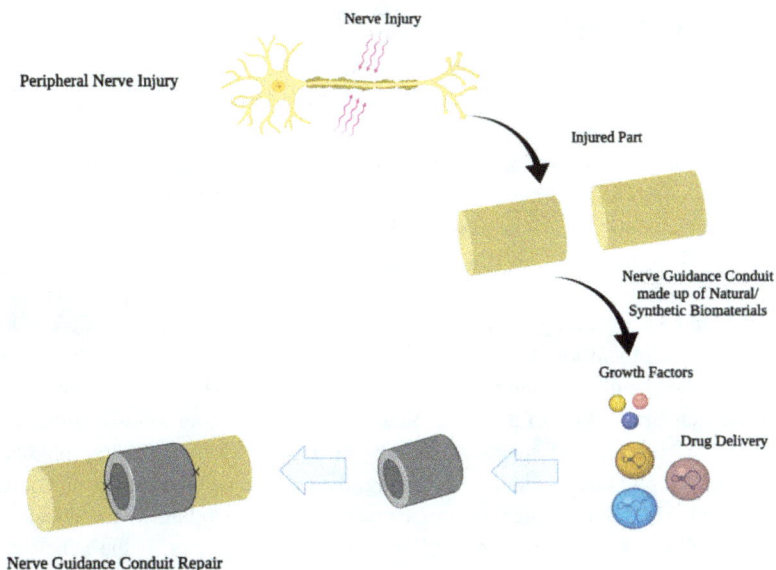

Figure 1: Schematic illustration of bridging a peripheral nerve injury defect through nerve guidance conduit to deliver growth factors and drugs to enhance the rate of healing

The proliferation, differentiation and attachment of cells are influenced by the topography and chemistry of the surface material. The uniform alignment of cells is favored by appropriate properties of surface. Whereas theregeneration of peripheral nervous system PNS is improved by the directional growth of axons. Additionally, NGCs lumen can also be used to introduce complex structures, such as NGCs filled with filaments, sponges, etc. Multichannel NGCs are used to improve the nerve repair functional outcomes. An effective method of additive manufacturing to fabricate NGCs is 3D printing that allows the desired architectural structures and surface permeability. This method also incorporates growth factors and cells into NGCs simultaneously. There is a great potential for 3D-printed NGCs made from biopolymers in precision medicine for nerve regeneration [46].

3.   Use of polymer-based surgical glues in the neurosurgery

3.1   Surgical glues

Surgical glues belong to the biomaterial class which may adhere as it applies to the tissues and makes the injuries or tissues get sealed [47,48]. Another advantage they provide is an aid in attachment of medical devices by the respective injured tissues and also aid in post-operation sealing to prevent any leakage by that opening. Preferably, surgical glue must have the strength to bind strongly. Moreover, it should be user-friendly decomposable biocompatible and showing elasticity. Which in turn brings compliance to the patient [49]. The basic principle followed by the surgical glues is bio-adhesion i.e the capacity to adhere for the required duration of time with the biological substrate [50]. Two interactions the chemical and the physiological interactions culminate into the basic procedure of adhesion possessed by surgical glues. The chemical interaction comprises of the bond formation which may be chemical in nature as well as the ionic bond which forms between the lines of the two substrates. One of those is a biological and other is an adhesive substrate which is due to the presence of functional groups in it [47]. The major three phases in that interaction are polymer swelling and wetting, the membrane of the mucosa, and chain or polymer interdiffusion. Which in turn causes chemical bonding in the chains that are entangled. Contrary to that physiological interaction comprises of the adhesion by the mechanisms which are physiologically associated for instance the process of clotting of blood [47]. Owing to these bondings and usage of the material for the manufacturing of surgical glues, it can be categorized as natural as well as synthetic surgical glues.

Particular challenges appear in the neurosurgery like limited accessibility, structures of neurons that are delicate and cerebrospinal fluid CSF leakage danger which may cause infections of CNS. Staples as well as the sutures have confined applications in case of surgeries of spines and crane, mostly in case of the setting of trauma where time plays a major role. Staples are used in surgery for closing incisions and similarly, sutures are used in surgery for holding up the tissues [51]. That is why the surgical bioadhesive is considered to be the best option due to the method of application. In fact that avoids trauma, gives positive results and also shows fixation which is stress distributed [52].

3.2   Fibrin glue

Fibrin is defined as the prime glue used in many surgeries widely. The mode of action is to start the procedure of clotting of blood. It is wrapped in an injector which is dual-mode and comprises of the two components mainly. The CaCl2, factor viii and fibrinogen make up the first component. Whereas the other component consists of the antifibrinolytic agent and thrombin. Plasma contains fibrinogen, which plays a vital role in the clotting of blood.

On application over wounds, fibrinogen is stimulated by the thrombin that afterward becomes a monomer of fibrin. Whereas Factor viii enables crosslinking of monomers of fibrin by enabling insoluble clot formation. Moreover, the role of calcium ions is to promote these biological processes [49,53,54].

During the surgery, fibrin glue behaves as an agent which is hemostatic in nature without the generation of heat. However, that seems to be expensive having poor mechanical power and mild property of adhesion. Therefore, fibrin glue is used in addition to sutures to compensate their poor mechanical strength. More considerably fibrin glue has a high potential risk in the transmission of pathogens that are blood-borne as sutures. In 1998 fibrin glue was approved firstly by the FDA. Commercially available examples of the fibrin glues include CrosseaL, Hemaseel and Crosseal [49,53,55].

Autologous fibrin sealant is an alternate to the already marketed available sealants. The process involved is cryoprecipitation in the preparation of sealant. Fibrinogen is attained by the blood banks or by the blood of the patient and has lower chances of transmission of pathogens that are blood-borne. The major side effect of using this fibrinogen is less availability of concentration of fibrinogen received compared to the already marketed sealants [55]. However, for the closing of wounds, fibrin glue is used to control bleeding and leak [56]. (see figure 2).

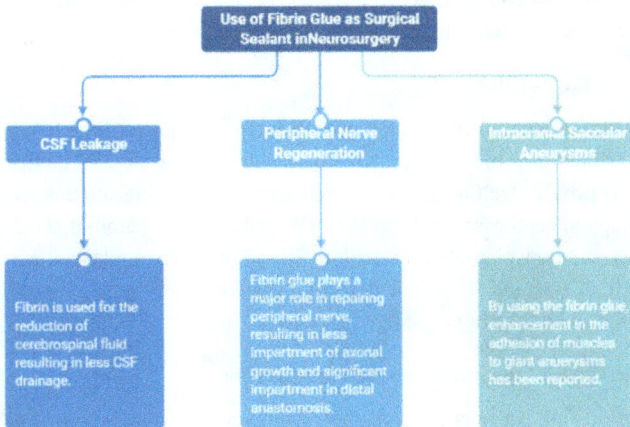

Figure 2: Use of fibrin glue as a bioadhesive in draining CSF leakage, repairing peripheral nerve injury, and intracranial saccular aneurysms

3.2.1   Fibrin glue in cerebrospinal fluid leakage

In dural closing, fibrin glue is widely used for the treatment and inhibition of the leakage of cerebrospinal fluid [57–59]. It is reported that the fibrin works for reduction of leakage of the cerebrospinal fluid after operation. Three groups were made of 39 patients. Autologous fibrin was used as sealent in one group for the treatment. Second group was treated with commercially available treatment and the third group was not treated with any fibrin sealant. On the third day, the patients treated with the fibrin sealant showed less drainage in cerebrospinal fluid but there was no significant difference observed between commercial and autologous [58,60]. It is concluded that the autologous sealant was found to be a cheaper and safe option compared to the commercial sealant. Other functions include control bleeding of the cranial tumor, fixation of nerve anastomoses, brain tumor embolization and hemangiomas [61].

When the carotid artery was ligated, fibrin glue is used for the treatment of the fistula of carotid-cavernous [62]. A commercially available sealant tends to minimize the cerebrospinal fluid leakage by having it sealed. Another study was done for the determination of the efficacy of tisseel clinically, on the individuals who have gone through the surgical procedure of the intracranial pathological lesions. Those individuals having chances of drainage in the CSF were given treatment and a comparison was made. It is concluded that individuals who got treatment with sealant mainly show no leakage. While those who got treatment with tisseel had 4-16 % leakage [63].

3.2.2   Fibrin glue as filler

Among the major objectives of surgery one is filling spaces, which hinders leakages, collection of fluids, aids in the healing of wounds and hemorrhages. The fibrin sealant (FS) has effectively proved itself for the purpose of sealing and filling and the treatment of initial level of leakage in cerebrospinal fluid [64]. For holes in the region of skull, the bone meal received from the transplantation is used as filler. Knoringer suggested a method in order to mask medium size and small size defects in the bones of the base of skull as well as a cap by mixing FS and the bone [65].

3.2.3   Fibrin glue in peripheral nerve regeneration

For the repairing of peripheral nerve, fibrin glue plays a major role. Egloff et al. [66] worked cases in which 23 cases were of brachial plexus 17 cases were of main nerve trunk where fibrin glue was applied at the site of repair for the sleeve. It is concluded that not much impartment of the growth of axonal but some impartment in distal anastomosis. The fibrin glue is also recommended as a remarkable suture for the purpose of repairing of nerves. Moreover, it is stated that it is required by 2/3rd times for the functional estimation

Applications of Polymers in Surgery
Materials Research Foundations **123** (2022) 92-122

Materials Research Forum LLC
https://doi.org/10.21741/9781644901892-4

of outcomes which analyze the axonal regeneration, anastomosis fibrosis and positioning of the fascicles mentioned [67].

3.2.4 Fibrin glue in intracranial saccular aneurysms

Intracranial saccular aneurysms develops due to the degeneration and developmental changes in the arteries of cerebral region. Owing to hypertension and a rise in the velocity of blood, a saccular aneurysm develops which can cause damage. It is proved that by using the fibrin glue, adhesion of muscles to giant aneurysms might be enhanced [68].

3.2.5 Fibrin glue in extra-intracranial anastomosis

Fibrin glue is also used as the extra intracranial anastomosis among the cortical branch and superficial temporal artery in order to inhibit ischemia in the cerebral region [69]. Moreover, for the defects of ventricles, closing of frontobasal fistulae, sealing the dural leaks resectioning the wall of sinus from outside, and sealing the leakage of subgalateal cerebrospinal fluid [70].

4. Use of polymers as dural substitutes

The defects in dura mater in the central nervous system may result from neurosurgical operations, erosions caused by tumors and severe trauma [71]. During the repair process of dura mater, it is required that leakage of cerebrospina fluid is prevented along with adhesion, intracranial infections and other complications. Over a centuary, many efforts have been put by scientists to create dural substitutes that are better to make the repair process improved and also to reduce asssociated complications. To avoid pathogens from spreading due to use of xenogenic or allogenic materials, researhers are trying to use several synthetic polymers as the substitute of dura mater that would not only reduce the cost but would also result in enhanced clinical efficacy [72].

After implantation of synthetic polymers, they can be sub-divided into two groups based on their degradation profile. These are absorbable and nonabsorbable materials. Absorable materials include the copolymer of L-lactic acid and epsilon-caprolactone, PGA, and copolymer of lactide and polydioxanone. Whereas the polyurethane, and expanded polytetrafluoroethylene dural substitites form the nonabsorbable materials. *In vivo* the absorbable materials face difficulty while degrading. That may result in friction with spinal cord or cerebral cortex, foreign body reactions and other complications associated with post operation [73–75].

Electrospining is used to stretch the polymers into filaments, and a structure is formed that is three-dimensional and naturally similar to dura structure. Cracks or holes are provided by that porous and bionic scaffold, which helps the autologous cells of the patients in migration. For the implanted material to degrade and the dura mater to reconstruct, that migrations of cells have great importance [76]. Additionally, adjustment can be made in the toughness of fibres when the concentration and composition of each polymer is changed. This also helps to obtain the required level of mechanical strength for the synthetic sbustitute of dura mater [77].

4.1   Polytetrafluoroethylene

When monomers of tetrafluoro ethylene combine, they form Polytetrafluoroethylene. The carbon skeleton is surrounded by the flourine atoms like a sheath. This gives this polymer great stability against number of chemical reactions. For dural repair, expanded polytetrafluoroethylene (EPTFE) which is a porous material isused. It is indicated when used *in vivo*, that EPTFE shows long-term durability without or very little adhesion or reaction to the living cells of the body. The EPTFE therefore, is considered as a suitable material to be used as an artificial dura mater because of these features [78].

4.2   Vicryl collagen

It is a bovine collagen which is coated with vicryl mesh or polyglactin 910. It is an adequate substitute of dura mater because it is safe, efficient and watertight. It also contains pleasent properties for surgeries. In less than 2 months, the material is reabsorbed and replaced with fibrous tissue with minimal adhesion to cortex or causing no or very little inflammation. When compared with other materials such as polyesterurethane and Lyodura during surgery, Vicryl ollagen is very good to handle becuase it is less rigid and during suturing, its tendency to tear is also less [79]. Since the material is expensive relatively, therefore it is recommended to use that only when the fibrous tissues of the patient are not sufficient enough in the operative field. The reabsorability and biocompatibility of Vicryl Collagen is its biggest advantage. But that also becomes a demerit in case of infections particularly when barrier of Vicryl Collagen is destroyed by the enhanced reaction of inflammation more rapidly than the growth of the fibrous tissues [80].

4.3   Elastin-fibrin material-neuroplast

The elastin-fibrin material is biodegradeable and the connective tissue matrix gradually replaces it. This connective matrix acts as a natural dura mater. Additionally, there is also a requirement for the artificial membrane to be flexible and strong. Indeed that should be capable of holding the edges of the substance that has lost by its tight binding . All these qualities and properties possessed by Elastin-fibrin material-neuroplast makes it a common choice for neurosurgery. That material has the ability to act as a reconstruction material and as a scaffold. After a period of time, that material disappears without causing any complications due to mutation or immunological actions. In fact that material is a good substitute of dura mater due to its compatibility, flexability and degradability [81].

5.   Use of polymers for incorporation of cells and growth factors

The delivery of bioactive molecules and exogenous cells to the injury site can ne modulated by the inflammatory response. Additionally, endogenous stem cells are also stimulated that promotes plasticity and neuroprotection. This is in the case of inury when regeneration capacity and endogenous neurogenesis is llimited. [82]. The cells which are transplanted may help in the process of repair of tissue by targeting host tissues or other factors which enhance neurogenesis [83].

Microenvironment can be provided by hydrogels and factors to enhance the integration of cells that are transplanted. Matrigel, a frequently used gel that is obtained from the sarcoma of the mouse with components of the ECM Collagen, laminin, entactin. That results in the reduction of size after damage, in the case when commonly used with the cells of transplantation [84]. With the ability to promote integration of neural cells and their protection for neural cell survival, further studies on the therapy of cells in addition to hydrogels protective in nature will significantly advance the engineering of neural tissue. Integration of factors of growth and other molecules that are bioactive biomaterials permit the scholars to bring factors that are specific for the site along with sequential control about the issue [85].

The dependency of release factor of material does not require the mechanical and chemical properties but depends on the encapsulation technique like covalent binding or loading directly. Distribution of vascular endothelial growth Factor by the PLGA microspheres enlist cells of endothelium and enhance the network progress of local neovascular [86]. Other factors of neurotrophic for instance GDND, BDNF and NGF are utilized in the animal model treatment and resulted in survival, transplanted cell differentiation and proliferation. In order to promote cell invasion, injected hydrogel vascularization is a vital thought. Tactics which involve biomaterial, growth factor and cells at present being used as the tool to enhance survival of cell and integration, also delivery of molecules that are bioactive at the site of damage of tissue. Cell integration and factor of growth as the vehicle into the biomaterial can offer support to cells and sustain release profile of the drug, therefore avoiding the necessity of various injections for delivery of drug [87].

6. Use of polymers as embolic material

For various neurovascular diseases such as spinal or cerebral arteriovenous malformations (AVMs), cerebral aneurysms,and dural arteriovenous fistulas, the strategy that is getting popular for their treatment is known as endovascular neurosurgery. In this procedure, a microcatheter is inserted into the lesion or close to it and to obliterate the lesion there is administration of an embolic material through the microcatheter. For embolization, various biomaterials are utilized. But each of the materials has its own problem. Poly(ethylene-co-vinyl alcohol) [88,89], a cationic methacrylate copolymer (Eudragit-E) [90,91], cellulose acetate [92,93] and poly(vinyl acetate) emulsion [94] are the polymer solutions that are used for embolization as liquid-type materials. Organic solvents such as ethanol and dimethyl sulfoxide are present in most of them.

When solution of a polymer is introduced into the blood vessel, the blood and surrounding tissues are diffused by that organic solvent. Thus there is a precipitation of polymeric substance to create an emboli. It is observed that surrounding tissues are affected in a toxic way by the ethanol [95,96] and dimethyl sulfoxide [97]. Moreover, microcatheters and their plastic hubs are also damaged by the dimethyl sulfoxide. Cyanoacrylate monomers are included in other types of embolic materials that are liquid in nature. It polymerizes instantly when it comes in contact with the blood. Among cyanoacrylate monomers, the most preferably used is *n*-butyl-2-cyanoacrylate (NBCA) [98–100].

6.1 *n*-butyl-2-cyanoacrylate (NBCA)

In1980s, *n*-butyl-2-cyanoacrylate (NBCA) was recognized as a material that may cause embolism. In 2000, FDA approved it to be used for the treatment of cerebral arteriovenous malformations [101,102]. When the NBCA is in its monomer form, that is a clear liquid. When it comes in conatct with any ionic substance like saline, blood, vascular endothelium and ionic contrast media, it becomes activated and polymerizes to form a matrix which is rigid [102]. Depending on the anatomy of target that need to be occluded and the flow characteristics of NBCA, 25 to 67% e concentration of NBCA is recommended. When ethiodized oil is combined with NBCA, the polymerization is slowed down in static blood. 67% NBCA slows down poymerization for less than 1 second and 25% NBCA slows down polymerization for 6 seconds [102]. Formaldehyde is released by the NBCA polymerization that leads to the vessel wall inflammation and the tissues of the surrounding. This causes chronic granulomatous inflammation eventually [100,102,103]. Moreover, before administration of NBCA, its mixture is prepared immediately becuase polymerization of NBCA starts when it is exposed to air [104].

7.   Polymers in stroke

One of the main reasons affecting a person's ability to lead a normal life is stroke. Every year almost 800,000 individuals suffer from it. Among them, ischemic stroke is observed in 80% of the cases [105]. It is also predicted that there will be a boom in stroke survivors carrying disabilities due to increasing accidents of strokes and the reduction in their rate of mortality [106] Limited achievements are made by the regenerative techniques such as stem cell therapy for betterment in restoration of normal behaviors in stroke patients. Unfortunately no replacement for misplaced tissues is found yet [107,108]. That is the reason behind the big tissue cavity which is present in the stroke survivors [109].

Scaffolds made from injectable biomaterials can be used to fill the cavity formed after stroke. The interaction between the host tissue and the transplanted material can be promoted [86,110–112]. The damaged tissue can be delivered with growth factors and drugs. The engraftment and the attachment of the transplanted cells can also be promoted as well as helping the lost cells to repopulate by recruiting the host cells is possible with bio-polymers [87].

The natural and synthetic polymers that are used to derive the protective hydrogels, can be used for the incorporation of different cells, growth factors and other medicinal agents. That is to enhance the microenvironment and also to release the bioactive material in a controlled manner. The natural or synthetic polymers in the 3D network are used to form the biomaterial scaffolds. An appropriate environment is provided by these scaffolds for cells to get attached, differentiate, and proliferate. The formation of ECM is facilitated by it [113]. It is vital to take note that the transplanted cells and their fate is determined by the mechanical and chemical properties of the biomaterial as well as the release profile of the drug. The chemical characterization of the biomaterial prepared directly affects the degradation rate and the extent to which molecules cross-link. The overall functionality is also determined for biomaterial through it [114].

7.1   Natural biomaterials used in the stroke

The 20% volume of entire tissues of the brain is made up of the extracellular matrix (ECM). Maintaining the key functions of the brain cells is an important role of ECM [115]. Hydrogels that are prepared using the ECM are responsible for the provision of mechanical properties. Along with that attraction of host cells through signaling molecules in the lession cavity is also provided by it. ECM derived hydrogels also obviate the requirement for exogenous cells [86,108,110,111]. Myelin, growth factors such as fibroblast, and VEGF, and ECM proteins such as fibronectin, laminin, etc., are present in decellularized biomaterial [116,117].

Hamethylcellulose (HAMC) [118], hyaluronic acid (HA) [119,120], chitosan [121,122], fibrin [123], and collagen [124] are natural polymers that have been used extensively beside ECM-derived hydrogels to deliver molecules or cells in the central nervous system. Collagen is the protein that makes up the main component of the extracellular matrix of peripheral tissues and also presents abundantly in mammals and that is why it is the most popular material which is used in biomedical applications. Several types of tissue engineering applications have made use of collagen hydrogels to encapsulate stem cells. This is because of collagen's mechanical strength, biocompatibility, and immunogenicity [125]. Hyaluronan is another polysaccharide that is present naturally in the central nervous system and it is also used for preparing hydrogels. It is known to promote cell survival and adhesion and shows the anti-inflammatory properties [126]. The Neural progenitor cells (NPC) survival are significantly promoted in mouse models of ischemia when transplantation is carried out of cross-linked heparin sulfate and hyaluronan hydrogel [127].

7.2   Synthetic biomaterials used in the stroke

Precise control is allowed over the properties of the material and the rates of degradation by the synthetic biomaterial by slowing down the release of drugs or small molecules in a controlled manner in the surrounding tissues. Polylactide (PL), the copolymers of lactide, and glycolide (PLGA) and polyglycolide (PG) are the polymeric agents that provide the controlled delivery of drugs and these are the most commonly used biomaterials. Bioactive molecules are loaded in particles of PLGA. These are then delivered to the injury site or for further tunning of location and delivery rate, these are embedded in hydrogels. Unlike natural biomaterials like Matrigel and collagen, synthetic biomaterials are defined better chemically. These are also inert biologically that reduce the chances of host immune response and variability [128].

The byproducts of synthetic polymers can be similar to natural biomaterials because of their bioactivity. The local microenvironment can also get influenced by them. It is suggested that the development of oligodendrocyte and myelination is supported by the application of lactic acid on the cultured slices of the developing cerebral cortex of the mouse. Lactic acid is the byproduct of PLGA [129]. For drug delivery, gels and nanoparticles are made from PLA, PGA, and PLGA because their rate of degradation can be controlled by controlling simply the ratio of PG:PL [130].

Poly (ethylene glycol) (PEG) is known as a synthetic polymer because of its property to resist absorption of protein and that is why it is used commonly in biomaterial applications. Although PEG hydrogels do not directly attach with the cells but often other polymers are mixed with it like gelatin or hyaluronic acid (HA) to support the attachment of cells and migration. A significant increase in the penetration of tissue and endogenous stimulation of NSC is observed with the epidermal growth factor modified with PEG (PEG-EGF) through epi-cortical delivery [131]. There are many promising reasons for the use of hydrogels based on PEG for cell and drug delivery, including nontoxicity, controlled degradation rate or rate of drug delivery as well as biocompatibility.

Summary

Polymers both natural and synthetic are being utilized for several reason in medicinal field. Their chemical, physical and biological properties make them suitable for use in many neurosurgery where conventional interventions do not help. Polymers have the possibility of nerve generation a reality. The different classes of polymers with their tailored formulations, co-polymerization, degree of crystallization, and blending with other polymers allow their application is a variety of neurosurgical procedures.

References

[1] D. Ozdil, H.M. Aydin, Polymers for medical and tissue engineering applications, Journal of Chemical Technology & Biotechnology. 89 (2014) 1793-1810. https://doi.org/10.1002/jctb.4505

[2] X. Liu, J.M. Holzwarth, P.X. Ma, Functionalized synthetic biodegradable polymer scaffolds for tissue engineering, Macromol Biosci. 12 (2012) 911-919. https://doi.org/10.1002/mabi.201100466

[3] Z.Z. Khaing, R.C. Thomas, S.A. Geissler, C.E. Schmidt, Advanced biomaterials for repairing the nervous system: what can hydrogels do for the brain?, Materials Today. 17 (2014) 332-340. https://doi.org/10.1016/j.mattod.2014.05.011

[4] J. Carmen, T. Magnus, R. Cassiani-Ingoni, L. Sherman, M.S. Rao, M.P. Mattson, Revisiting the astrocyte-oligodendrocyte relationship in the adult CNS, Progress in Neurobiology. 82 (2007) 151-162. https://doi.org/10.1016/j.pneurobio.2007.03.001

[5] H. Zhang, K. Uchimura, K. Kadomatsu, Brain Keratan Sulfate and Glial Scar Formation, Annals of the New York Academy of Sciences. 1086 (2006) 81-90. https://doi.org/10.1196/annals.1377.014

[6] R.G. Ellis-Behnke, L.A. Teather, G.E. Schneider, K.F. So, Using Nanotechnology to Design Potential Therapies for CNS Regeneration, Current Pharmaceutical Design. 13 (2007) 2519-2528. https://doi.org/10.2174/138161207781368648

[7] M. Tsintou, C. Wang, K. Dalamagkas, D. Weng, Y.-N. Zhang, W. Niu, 5 - Nanogels for biomedical applications: Drug delivery, imaging, tissue engineering, and

biosensors, in: M. Razavi, A. Thakor (Eds.), Nanobiomaterials Science, Development and Evaluation, Woodhead Publishing, 2017: pp. 87-124. https://doi.org/10.1016/B978-0-08-100963-5.00005-7

[8] A. Fraczek-Szczypta, Carbon nanomaterials for nerve tissue stimulation and regeneration, Materials Science and Engineering: C. 34 (2014) 35-49. https://doi.org/10.1016/j.msec.2013.09.038

[9] S. Verma, A.J. Domb, N. Kumar, Nanomaterials for regenerative medicine, Nanomedicine (Lond). 6 (2011) 157-181. https://doi.org/10.2217/nnm.10.146

[10] O.S. Manoukian, J.T. Baker, S. Rudraiah, M.R. Arul, A.T. Vella, A.J. Domb, S.G. Kumbar, Functional polymeric nerve guidance conduits and drug delivery strategies for peripheral nerve repair and regeneration, Journal of Controlled Release. 317 (2020) 78-95. https://doi.org/10.1016/j.jconrel.2019.11.021

[11] J. Noble, C.A. Munro, V.S. Prasad, R. Midha, Analysis of upper and lower extremity peripheral nerve injuries in a population of patients with multiple injuries, J Trauma. 45 (1998) 116-122. https://doi.org/10.1097/00005373-199807000-00025

[12] S. Yegiyants, D. Dayicioglu, G. Kardashian, Z.J. Panthaki, Traumatic peripheral nerve injury: a wartime review, J Craniofac Surg. 21 (2010) 998-1001. https://doi.org/10.1097/SCS.0b013e3181e17aef

[13] S.E. Mackinnon, A.R. Hudson, R.E. Falk, D. Kline, D. Hunter, Peripheral nerve allograft: an assessment of regeneration across pretreated nerve allografts, Neurosurgery. 15 (1984) 690-693. https://doi.org/10.1227/00006123-198411000-00009

[14] P.J. Apel, J. Ma, M. Callahan, C.N. Northam, T.B. Alton, W.E. Sonntag, Z. Li, Effect of locally delivered IGF-1 on nerve regeneration during aging: an experimental study in rats, Muscle Nerve. 41 (2010) 335-341. https://doi.org/10.1002/mus.21485

[15] P. Sierpinski, J. Garrett, J. Ma, P. Apel, D. Klorig, T. Smith, L.A. Koman, A. Atala, M. Van Dyke, The use of keratin biomaterials derived from human hair for the promotion of rapid regeneration of peripheral nerves, Biomaterials. 29 (2008) 118-128. https://doi.org/10.1016/j.biomaterials.2007.08.023

[16] S. Song, X. Wang, T. Wang, Q. Yu, Z. Hou, Z. Zhu, R. Li, Additive Manufacturing of Nerve Guidance Conduits for Regeneration of Injured Peripheral Nerves, Front. Bioeng. Biotechnol. 8 (2020). https://doi.org/10.3389/fbioe.2020.590596

[17] R. Deumens, A. Bozkurt, M.F. Meek, M.A.E. Marcus, E.A.J. Joosten, J. Weis, G.A. Brook, Repairing injured peripheral nerves: Bridging the gap, Prog Neurobiol. 92 (2010) 245-276. https://doi.org/10.1016/j.pneurobio.2010.10.002

[18] X. Zhang, W. Qu, D. Li, K. Shi, R. Li, Y. Han, E. Jin, J. Ding, X. Chen, Functional Polymer-Based Nerve Guide Conduits to Promote Peripheral Nerve Regeneration, Advanced Materials Interfaces. 7 (2020) 2000225. https://doi.org/10.1002/admi.202000225

[19] K.Y. Lee, D.J. Mooney, Alginate: properties and biomedical applications, Prog Polym Sci. 37 (2012) 106-126. https://doi.org/10.1016/j.progpolymsci.2011.06.003

[20] B.N. Johnson, K.Z. Lancaster, G. Zhen, J. He, M.K. Gupta, Y.L. Kong, E.A. Engel, K.D. Krick, A. Ju, F. Meng, L.W. Enquist, X. Jia, M.C. McAlpine, 3D Printed Anatomical Nerve Regeneration Pathways, Adv Funct Mater. 25 (2015) 6205-6217. https://doi.org/10.1002/adfm.201501760

[21] S. Suri, L.-H. Han, W. Zhang, A. Singh, S. Chen, C.E. Schmidt, Solid freeform fabrication of designer scaffolds of hyaluronic acid for nerve tissue engineering, Biomed Microdevices. 13 (2011) 983-993. https://doi.org/10.1007/s10544-011-9568-9

[22] B. Weng, X. Liu, R. Shepherd, G.G. Wallace, Inkjet printed polypyrrole/collagen scaffold: A combination of spatial control and electrical stimulation of PC12 cells, Synthetic Metals. 162 (2012) 1375-1380. https://doi.org/10.1016/j.synthmet.2012.05.022

[23] S.H. Kim, Y.K. Yeon, J.M. Lee, J.R. Chao, Y.J. Lee, Y.B. Seo, M.T. Sultan, O.J. Lee, J.S. Lee, S.-I. Yoon, I.-S. Hong, G. Khang, S.J. Lee, J.J. Yoo, C.H. Park, Precisely printable and biocompatible silk fibroin bioink for digital light processing 3D printing, Nat Commun. 9 (2018) 1620. https://doi.org/10.1038/s41467-018-03759-y

[24] Y. Hu, Y. Wu, Z. Gou, J. Tao, J. Zhang, Q. Liu, T. Kang, S. Jiang, S. Huang, J. He, S. Chen, Y. Du, M. Gou, 3D-engineering of Cellularized Conduits for Peripheral Nerve Regeneration, Sci Rep. 6 (2016) 32184. https://doi.org/10.1038/srep32184

[25] K. Ikeda, D. Yamauchi, N. Osamura, N. Hagiwara, K. Tomita, Hyaluronic acid prevents peripheral nerve adhesion, British Journal of Plastic Surgery. 56 (2003) 342-347. https://doi.org/10.1016/S0007-1226(03)00197-8

[26] M. Georgiou, J.P. Golding, A.J. Loughlin, P.J. Kingham, J.B. Phillips, Engineered neural tissue with aligned, differentiated adipose-derived stem cells promotes

peripheral nerve regeneration across a critical sized defect in rat sciatic nerve, Biomaterials. 37 (2015) 242-251. https://doi.org/10.1016/j.biomaterials.2014.10.009

[27] A. Bozkurt, A. Boecker, J. Tank, H. Altinova, R. Deumens, C. Dabhi, R. Tolba, J. Weis, G.A. Brook, N. Pallua, S.G.A. van Neerven, Efficient bridging of 20 mm rat sciatic nerve lesions with a longitudinally micro-structured collagen scaffold, Biomaterials. 75 (2016) 112-122. https://doi.org/10.1016/j.biomaterials.2015.10.009

[28] S. Kehoe, X.F. Zhang, D. Boyd, FDA approved guidance conduits and wraps for peripheral nerve injury: a review of materials and efficacy, Injury. 43 (2012) 553-572. https://doi.org/10.1016/j.injury.2010.12.030

[29] M.D. Sarker, S. Naghieh, A.D. McInnes, D.J. Schreyer, X. Chen, Regeneration of peripheral nerves by nerve guidance conduits: Influence of design, biopolymers, cells, growth factors, and physical stimuli, Prog Neurobiol. 171 (2018) 125-150. https://doi.org/10.1016/j.pneurobio.2018.07.002

[30] W. Zhu, K.R. Tringale, S.A. Woller, S. You, S. Johnson, H. Shen, J. Schimelman, M. Whitney, J. Steinauer, W. Xu, T.L. Yaksh, Q.T. Nguyen, S. Chen, Rapid continuous 3D printing of customizable peripheral nerve guidance conduits, Mater Today (Kidlington). 21 (2018) 951-959. https://doi.org/10.1016/j.mattod.2018.04.001

[31] A. Singh, S. Asikainen, A.K. Teotia, P.A. Shiekh, E. Huotilainen, I. Qayoom, J. Partanen, J. Seppälä, A. Kumar, Biomimetic Photocurable Three-Dimensional Printed Nerve Guidance Channels with Aligned Cryomatrix Lumen for Peripheral Nerve Regeneration, ACS Appl Mater Interfaces. 10 (2018) 43327-43342. https://doi.org/10.1021/acsami.8b11677

[32] C.J. Pateman, A.J. Harding, A. Glen, C.S. Taylor, C.R. Christmas, P.P. Robinson, S. Rimmer, F.M. Boissonade, F. Claeyssens, J.W. Haycock, Nerve guides manufactured from photocurable polymers to aid peripheral nerve repair, Biomaterials. 49 (2015) 77-89. https://doi.org/10.1016/j.biomaterials.2015.01.055

[33] M.S. Evangelista, M. Perez, A.A. Salibian, J.M. Hassan, S. Darcy, K.Z. Paydar, R.B. Wicker, K. Arcaute, B.K. Mann, G.R.D. Evans, Single-lumen and multi-lumen poly(ethylene glycol) nerve conduits fabricated by stereolithography for peripheral nerve regeneration in vivo, J Reconstr Microsurg. 31 (2015) 327-335. https://doi.org/10.1055/s-0034-1395415

[34] D. Radulescu, S. Dhar, C.M. Young, D.W. Taylor, H.-J. Trost, D.J. Hayes, G.R. Evans, Tissue engineering scaffolds for nerve regeneration manufactured by ink-jet technology, Materials Science and Engineering: C. 27 (2007) 534-539. https://doi.org/10.1016/j.msec.2006.05.050

[35] D. Singh, A.J. Harding, E. Albadawi, F.M. Boissonade, J.W. Haycock, F. Claeyssens, Additive manufactured biodegradable poly(glycerol sebacate methacrylate) nerve guidance conduits, Acta Biomater. 78 (2018) 48-63. https://doi.org/10.1016/j.actbio.2018.07.055

[36] S.-J. Lee, T. Esworthy, S. Stake, S. Miao, Y.Y. Zuo, B.T. Harris, L.G. Zhang, Advances in 3D Bioprinting for Neural Tissue Engineering, Advanced Biosystems. 2 (2018) 1700213. https://doi.org/10.1002/adbi.201700213

[37] Y.D. Kim, J.H. Kim, Synthesis of polypyrrole-polycaprolactone composites by emulsion polymerization and the electrorheological behavior of their suspensions, Colloid Polym Sci. 286 (2008) 631-637. https://doi.org/10.1007/s00396-007-1802-x

[38] E. Schnell, K. Klinkhammer, S. Balzer, G. Brook, D. Klee, P. Dalton, J. Mey, Guidance of glial cell migration and axonal growth on electrospun nanofibers of poly-epsilon-caprolactone and a collagen/poly-epsilon-caprolactone blend, Biomaterials. 28 (2007) 3012-3025. https://doi.org/10.1016/j.biomaterials.2007.03.009

[39] W. Chang, M.B. Shah, P. Lee, X. Yu, Tissue-engineered spiral nerve guidance conduit for peripheral nerve regeneration, Acta Biomater. 73 (2018) 302-311. https://doi.org/10.1016/j.actbio.2018.04.046

[40] L. Huang, L. Zhu, X. Shi, B. Xia, Z. Liu, S. Zhu, Y. Yang, T. Ma, P. Cheng, K. Luo, J. Huang, Z. Luo, A compound scaffold with uniform longitudinally oriented guidance cues and a porous sheath promotes peripheral nerve regeneration in vivo, Acta Biomater. 68 (2018) 223-236. https://doi.org/10.1016/j.actbio.2017.12.010

[41] S.H. Oh, J.H. Kim, K.S. Song, B.H. Jeon, J.H. Yoon, T.B. Seo, U. Namgung, I.W. Lee, J.H. Lee, Peripheral nerve regeneration within an asymmetrically porous PLGA/Pluronic F127 nerve guide conduit, Biomaterials. 29 (2008) 1601-1609. https://doi.org/10.1016/j.biomaterials.2007.11.036

[42] C. Xue, N. Hu, Y. Gu, Y. Yang, Y. Liu, J. Liu, F. Ding, X. Gu, Joint use of a chitosan/PLGA scaffold and MSCs to bridge an extra large gap in dog sciatic nerve, Neurorehabil Neural Repair. 26 (2012) 96-106. https://doi.org/10.1177/1545968311420444

[43] K. Arcaute, B.K. Mann, R.B. Wicker, Stereolithography of three-dimensional bioactive poly(ethylene glycol) constructs with encapsulated cells, Ann Biomed Eng. 34 (2006) 1429-1441. https://doi.org/10.1007/s10439-006-9156-y

[44] Ş.U. Karabiberoğlu, Ç.C. Koçak, Z. Dursun, Carbon Nanotube-Conducting Polymer Composites as Electrode Material in Electroanalytical Applications, IntechOpen, 2016. https://doi.org/10.5772/62882

[45] J. Connor, D. McQuillan, M. Sandor, H. Wan, J. Lombardi, N. Bachrach, J. Harper, H. Xu, Retention of structural and biochemical integrity in a biological mesh supports tissue remodeling in a primate abdominal wall model, Regen Med. 4 (2009) 185-195. https://doi.org/10.2217/17460751.4.2.185

[46] Y. Liu, S. Hsu, Biomaterials and neural regeneration, Neural Regen Res. 15 (2020) 1243-1244. https://doi.org/10.4103/1673-5374.272573

[47] M.L.B. Palacio, B. Bhushan, Bioadhesion: a review of concepts and applications, Philosophical Transactions of the Royal Society A: Mathematical, Physical and Engineering Sciences. 370 (2012) 2321-2347. https://doi.org/10.1098/rsta.2011.0483

[48] M. Kazemzadeh-Narbat, N. Annabi, A. Khademhosseini, Surgical sealants and high strength adhesives, Materials Today. 18 (n.d.) 176-177. https://doi.org/10.1016/j.mattod.2015.02.012

[49] H.L. Mobley, R.M. Garner, G.R. Chippendale, J.V. Gilbert, A.V. Kane, A.G. Plaut, Role of Hpn and NixA of Helicobacter pylori in susceptibility and resistance to bismuth and other metal ions, Helicobacter. 4 (1999) 162-169. https://doi.org/10.1046/j.1523-5378.1999.99286.x

[50] F. Scognamiglio, A. Travan, I. Rustighi, P. Tarchi, S. Palmisano, E. Marsich, M. Borgogna, I. Donati, N. de Manzini, S. Paoletti, Adhesive and sealant interfaces for general surgery applications, J Biomed Mater Res B Appl Biomater. 104 (2016) 626-639. https://doi.org/10.1002/jbm.b.33409

[51] T.O. Smith, D. Sexton, C. Mann, S. Donell, Sutures versus staples for skin closure in orthopaedic surgery: meta-analysis, BMJ. 340 (2010) c1199. https://doi.org/10.1136/bmj.c1199

[52] L. Qiu, A.A. Qi See, T.W.J. Steele, N.K. Kam King, Bioadhesives in neurosurgery: a review, J Neurosurg. (2019) 1-11.

[53] N. Annabi, K. Yue, A. Tamayol, A. Khademhosseini, Elastic sealants for surgical applications, Eur J Pharm Biopharm. 95 (2015) 27-39. https://doi.org/10.1016/j.ejpb.2015.05.022

[54] W.D. Spotnitz, Commercial fibrin sealants in surgical care, Am J Surg. 182 (2001) 8S-14S. https://doi.org/10.1016/S0002-9610(01)00771-1

[55] L. Foster, Bioadhesives as Surgical Sealants: A Review, in: 2015: pp. 203-234. https://doi.org/10.1201/b18095-14

[56] G. Lopezcarasa-Hernandez, J.-F. Perez-Vazquez, J.-L. Guerrero-Naranjo, M.A. Martinez-Castellanos, Versatility of use of fibrin glue in wound closure and vitreo-retinal surgery, International Journal of Retina and Vitreous. 7 (2021) 33. https://doi.org/10.1186/s40942-021-00298-5

[57] D.M. Albala, Fibrin sealants in clinical practice, Cardiovasc Surg. 11 Suppl 1 (2003) 5-11. https://doi.org/10.1016/S0967-2109(03)00065-6

[58] H. Nakamura, Y. Matsuyama, H. Yoshihara, Y. Sakai, Y. Katayama, S. Nakashima, J. Takamatsu, N. Ishiguro, The effect of autologous fibrin tissue adhesive on postoperative cerebrospinal fluid leak in spinal cord surgery: a randomized controlled trial, Spine (Phila Pa 1976). 30 (2005) E347-351. https://doi.org/10.1097/01.brs.0000167820.54413.8e

[59] S. Ito, K. Nagayama, N. Iino, I. Saito, Y. Takami, Frontal sinus repair with free autologous bone grafts and fibrin glue, Surgical Neurology. 60 (2003) 155-158. https://doi.org/10.1016/S0090-3019(03)00207-6

[60] D.M. Albala, J.H. Lawson, Recent clinical and investigational applications of fibrin sealant in selected surgical specialties, J Am Coll Surg. 202 (2006) 685-697. https://doi.org/10.1016/j.jamcollsurg.2005.11.027

[61] H.A. Crockard, The transoral approach to the base of the brain and upper cervical cord, Ann R Coll Surg Engl. 67 (1985) 321-325.

[62] H. Hasegawa, S. Bitoh, J. Obashi, M. Maruno, Closure of carotid-cavernous fistulae by use of a fibrin adhesive system, Surgical Neurology. 24 (1985) 23-26. https://doi.org/10.1016/0090-3019(85)90057-6

[63] A. Kassam, M. Horowitz, R. Carrau, C. Snyderman, W. Welch, B. Hirsch, Y.F. Chang, Use of Tisseel fibrin sealant in neurosurgical procedures: incidence of cerebrospinal fluid leaks and cost-benefit analysis in a retrospective study, Neurosurgery. 52 (2003) 1102-1105; discussion 1105. https://doi.org/10.1227/01.NEU.0000057699.37541.76

[64] N.P. Biscola, L.P. Cartarozzi, S. Ulian-Benitez, R. Barbizan, M.V. Castro, A.B. Spejo, R.S. Ferreira, B. Barraviera, A.L.R. Oliveira, Multiple uses of fibrin sealant for nervous system treatment following injury and disease, Journal of Venomous Animals and Toxins Including Tropical Diseases. 23 (2017) 13. https://doi.org/10.1186/s40409-017-0103-1

[65] F.H. Silver, M.C. Wang, G.D. Pins, Preparation and use of fibrin glue in surgery, Biomaterials. 16 (1995) 891-903. https://doi.org/10.1016/0142-9612(95)93113-R

[66] D.V. Egloff, A. Narakas, C. Bonnard, Results of Nerve Grafts with Tissucol (Tisseel) Anastomosis, in: G. Schlag, H. Redl (Eds.), Fibrin Sealant in Operative Medicine, Springer, Berlin, Heidelberg, 1986: pp. 181-185. https://doi.org/10.1007/978-3-642-71391-0_26

[67] A. Narakas, The Use of Fibrin Glue in Repair of Peripheral Nerves, Orthopedic Clinics of North America. 19 (1988) 187-199. https://doi.org/10.1016/S0030-5898(20)30342-4

[68] J. Haase, E. Falk, F.F. Madsen, P.S. Teglbjærg, Experimental and Clinical Use of Tisseel (Tissucol) and Muscle to Reinforce Rat Arteries and Human Cerebral Saccular Aneurysms, in: G. Schlag, H. Redl (Eds.), Fibrin Sealant in Operative Medicine, Springer, Berlin, Heidelberg, 1986: pp. 172-175. https://doi.org/10.1007/978-3-642-71391-0_24

[69] G. Parenti, B. Lenzi, Clinical and Angiographic Results in 47 Microsurgical Anastomoses for Cerebral Revascularization, in: G. Schlag, H. Redl (Eds.), Fibrin Sealant in Operative Medicine, Springer, Berlin, Heidelberg, 1986: pp. 123-128. https://doi.org/10.1007/978-3-642-71391-0_17

[70] H.R. Eggert, W. Seeger, V. Kallmeyer, Application of Fibrin Sealant in Microneurosurgery, in: G. Schlag, H. Redl (Eds.), Fibrin Sealant in Operative Medicine, Springer, Berlin, Heidelberg, 1986: pp. 139-143. https://doi.org/10.1007/978-3-642-71391-0_19

[71] C.M. Poser, Trauma to the central nervous system may result in formation or enlargement of multiple sclerosis plaques, Arch Neurol. 57 (2000) 1074-1077, discussion 1078. https://doi.org/10.1001/archneur.57.7.1074

[72] K. Kurpinski, S. Patel, Dura mater regeneration with a novel synthetic, bilayered nanofibrous dural substitute: an experimental study, Https://Doi.Org/10.2217/Nnm.10.132. (2011). https://doi.org/10.2217/nnm.10.132

[73] M. Pierson, P.V. Birinyi, S. Bhimireddy, J.R. Coppens, Analysis of Decompressive Craniectomies with Subsequent Cranioplasties in the Presence of Collagen Matrix Dural Substitute and Polytetrafluoroethylene as an Adhesion Preventative Material, World Neurosurgery. 86 (2016) 153-160. https://doi.org/10.1016/j.wneu.2015.09.078

[74] N.-X. Xiong, D.A. Tan, P. Fu, Y.-Z. Huang, S. Tong, H. Yu, Healing of Deep Wound Infection without Removal of Non-Absorbable Dura Mater (Neuro-Patch®): A

Applications of Polymers in Surgery   Materials Research Forum LLC
Materials Research Foundations **123** (2022) 92-122   https://doi.org/10.21741/9781644901892-4

Case Report, J Long Term Eff Med Implants. 26 (2016) 43-48.
https://doi.org/10.1615/JLongTermEffMedImplants.2016010104

[75] Y.-H. Huang, T.-C. Lee, W.-F. Chen, Y.-M. Wang, Safety of the nonabsorbable dural substitute in decompressive craniectomy for severe traumatic brain injury, J Trauma. 71 (2011) 533-537. https://doi.org/10.1097/TA.0b013e318203208a

[76] K. Deng, Y. Yang, Y. Ke, C. Luo, M. Liu, Y. Deng, Q. Tian, Y. Yuan, T. Yuan, T. Xu, A novel biomimetic composite substitute of PLLA/gelatin nanofiber membrane for dura repairing, Neurol Res. 39 (2017) 819-829. https://doi.org/10.1080/01616412.2017.1348680

[77] P. Schmalz, C. Griessenauer, C.S. Ogilvy, A.J. Thomas, Use of an Absorbable Synthetic Polymer Dural Substitute for Repair of Dural Defects: A Technical Note, Cureus. 10 (n.d.) e2127.

[78] S. Yamagata, K. Goto, Y. Oda, H. Kikuchi, Clinical experience with expanded polytetrafluoroethylene sheet used as an artificial dura mater, Neurol Med Chir (Tokyo). 33 (1993) 582-585. https://doi.org/10.2176/nmc.33.582

[79] J.-S. Raul, J. Godard, F. Arbez-Gindre, A. Czorny, [Use of polyester urethane (Neuro-Patch) as a dural substitute. Prospective study of 70 cases], Neurochirurgie. 49 (2003) 83-89.

[80] F. Van Calenbergh, E. Quintens, R. Sciot, J. Van Loon, J. Goffin, C. Plets, The use of Vicryl Collagen® as a dura substitute: A clinical review of 78 surgical cases, Acta Neurochir. 139 (1997) 120-123. https://doi.org/10.1007/BF02747191

[81] F. San-Galli, G. Deminière, J. Guérin, M. Rabaud, Use of a biodegradable elastin-fibrin material, Neuroplast®, as a dural substitute, Biomaterials. 17 (1996) 1081-1085. https://doi.org/10.1016/0142-9612(96)85908-4

[82] P. Dibajnia, C.M. Morshead, Role of neural precursor cells in promoting repair following stroke, Acta Pharmacol Sin. 34 (2013) 78-90. https://doi.org/10.1038/aps.2012.107

[83] T.M. Bliss, R.H. Andres, G.K. Steinberg, Optimizing the Success of Cell Transplantation Therapy for Stroke, Neurobiol Dis. 37 (2010) 275. https://doi.org/10.1016/j.nbd.2009.10.003

[84] K. Jin, X. Mao, L. Xie, V. Galvan, B. Lai, Y. Wang, O. Gorostiza, X. Wang, D.A. Greenberg, Transplantation of human neural precursor cells in Matrigel scaffolding improves outcome from focal cerebral ischemia after delayed postischemic treatment

in rats, J Cereb Blood Flow Metab. 30 (2010) 534-544.
https://doi.org/10.1038/jcbfm.2009.219

[85] F.-M. Chen, X. Liu, Advancing biomaterials of human origin for tissue engineering, Prog Polym Sci. 53 (2016) 86-168. https://doi.org/10.1016/j.progpolymsci.2015.02.004

[86] E. Bible, O. Qutachi, D.Y.S. Chau, M.R. Alexander, K.M. Shakesheff, M. Modo, Neo-vascularization of the stroke cavity by implantation of human neural stem cells on VEGF-releasing PLGA microparticles, Biomaterials. 33 (2012) 7435-7446. https://doi.org/10.1016/j.biomaterials.2012.06.085

[87] H. Ghuman, M. Modo, Biomaterial applications in neural therapy and repair, Chinese Neurosurgical Journal. 2 (2016) 34. https://doi.org/10.1186/s41016-016-0057-0

[88] W. Taki, Y. Yonekawa, H. Iwata, A. Uno, K. Yamashita, H. Amemiya, A new liquid material for embolization of arteriovenous malformations, AJNR Am J Neuroradiol. 11 (1990) 163-168.

[89] K. Yamashita, W. Taki, H. Iwata, I. Nakahara, S. Nishi, A. Sadato, K. Matsumoto, H. Kikuchi, Characteristics of ethylene vinyl alcohol copolymer (EVAL) mixtures, AJNR Am J Neuroradiol. 15 (1994) 1103-1105.

[90] K. Yamashita, W. Taki, H. Iwata, H. Kikuchi, A cationic polymer, Eudragit-E, as a new liquid embolic material for arteriovenous malformations, Neuroradiology. 38 Suppl 1 (1996) S151-156. https://doi.org/10.1007/BF02278144

[91] A. Sadato, Y. Numaguchi, W. Taki, H. Iwata, K. Yamshita, Nonadhesive liquid embolic agent: role of its components in histologic changes in embolized arteries, Acad Radiol. 5 (1998) 198-206. https://doi.org/10.1016/S1076-6332(98)80284-5

[92] S. Mandai, K. Kinugasa, T. Ohmoto, Direct thrombosis of aneurysms with cellulose acetate polymer. Part I: Results of thrombosis in experimental aneurysms, J Neurosurg. 77 (1992) 497-500. https://doi.org/10.3171/jns.1992.77.4.0497

[93] K. Kinugasa, S. Mandai, Y. Terai, I. Kamata, K. Sugiu, T. Ohmoto, A. Nishimoto, Direct thrombosis of aneurysms with cellulose acetate polymer. Part II: Preliminary clinical experience, J Neurosurg. 77 (1992) 501-507. https://doi.org/10.3171/jns.1992.77.4.0501

[94] A. Sadato, W. Taki, Y. Ikada, I. Nakahara, K. Yamashita, K. Matsumoto, M. Tanaka, H. Kikuchi, Y. Doi, T. Noguchi, Experimental study and clinical use of

poly(vinyl acetate) emulsion as liquid embolisation material, Neuroradiology. 36 (1994) 634-641. https://doi.org/10.1007/BF00600429

[95] K. Buchta, J. Sands, H. Rosenkrantz, W.D. Roche, Early mechanism of action of arterially infused alcohol U.S.P. in renal devitalization, Radiology. 145 (1982) 45-48. https://doi.org/10.1148/radiology.145.1.7122894

[96] B.A. Ellman, B.J. Parkhill, P.B. Marcus, T.S. Curry, P.C. Peters, Renal ablation with absolute ethanol. Mechanism of action, Invest Radiol. 19 (1984) 416-423. https://doi.org/10.1097/00004424-198409000-00014

[97] J.C. Chaloupka, F. Viñuela, H.V. Vinters, J. Robert, Technical feasibility and histopathologic studies of ethylene vinyl copolymer (EVAL) using a swine endovascular embolization model, AJNR Am J Neuroradiol. 15 (1994) 1107-1115.

[98] D. Samson, Q.M. Ditmore, C.W. Beyer, Intravascular use of isobutyl 2-cyanoacrylate: Part 1 Treatment of intracranial arteriovenous malformations, Neurosurgery. 8 (1981) 43-51. https://doi.org/10.1097/00006123-198101000-00009

[99] G. Debrun, F. Vinuela, A. Fox, C.G. Drake, Embolization of cerebral arteriovenous malformations with bucrylate: Experience in 46 cases, Journal of Neurosurgery. 56 (1982) 615-627. https://doi.org/10.3171/jns.1982.56.5.0615

[100] M.F. Brothers, J.C. Kaufmann, A.J. Fox, J.P. Deveikis, n-Butyl 2-cyanoacrylate--substitute for IBCA in interventional neuroradiology: histopathologic and polymerization time studies, AJNR Am J Neuroradiol. 10 (1989) 777-786.

[101] R.E. Kania, E. Sauvaget, J.-P. Guichard, R. Chapot, P.T.B. Huy, P. Herman, Early postoperative CT scanning for juvenile nasopharyngeal angiofibroma: detection of residual disease, AJNR Am J Neuroradiol. 26 (2005) 82-88.

[102] R.J. Rosen, S. Contractor, The use of cyanoacrylate adhesives in the management of congenital vascular malformations, Semin Intervent Radiol. 21 (2004) 59-66. https://doi.org/10.1055/s-2004-831406

[103] H.V. Vinters, K.A. Galil, M.J. Lundie, J.C.E. Kaufmann, The histotoxicity of cyanoacrylates, Neuroradiology. 27 (1985) 279-291. https://doi.org/10.1007/BF00339559

[104] H. Hill, J.F.B. Chick, A. Hage, R.N. Srinivasa, N-butyl cyanoacrylate embolotherapy: techniques, complications, and management, Diagn Interv Radiol. 24 (2018) 98-103. https://doi.org/10.5152/dir.2018.17432

[105] A.G. Thrift, H.M. Dewey, R.A. Macdonell, J.J. McNeil, G.A. Donnan, Incidence of the major stroke subtypes: initial findings from the North East Melbourne stroke

incidence study (NEMESIS), Stroke. 32 (2001) 1732-1738.
https://doi.org/10.1161/01.STR.32.8.1732

[106] J.P. Broderick, William M. Feinberg Lecture: Stroke Therapy in the Year 2025, Stroke. 35 (2004) 205-211. https://doi.org/10.1161/01.STR.0000106160.34316.19

[107] E.J. Smith, R.P. Stroemer, N. Gorenkova, M. Nakajima, W.R. Crum, E. Tang, L. Stevanato, J.D. Sinden, M. Modo, Implantation site and lesion topology determine efficacy of a human neural stem cell line in a rat model of chronic stroke, Stem Cells. 30 (2012) 785-796. https://doi.org/10.1002/stem.1024

[108] A. Encarnacion, N. Horie, H. Keren-Gill, T.M. Bliss, G.K. Steinberg, M. Shamloo, Long-term behavioral assessment of function in an experimental model for ischemic stroke, J Neurosci Methods. 196 (2011) 247-257.
https://doi.org/10.1016/j.jneumeth.2011.01.010

[109] F. Moreau, S. Patel, M.L. Lauzon, C.R. McCreary, M. Goyal, R. Frayne, A.M. Demchuk, S.B. Coutts, E.E. Smith, Cavitation after acute symptomatic lacunar stroke depends on time, location, and MRI sequence, Stroke. 43 (2012) 1837-1842.
https://doi.org/10.1161/STROKEAHA.111.647859

[110] K.I. Park, Y.D. Teng, E.Y. Snyder, The injured brain interacts reciprocally with neural stem cells supported by scaffolds to reconstitute lost tissue, Nat Biotechnol. 20 (2002) 1111-1117. https://doi.org/10.1038/nbt751

[111] E. Bible, D.Y.S. Chau, M.R. Alexander, J. Price, K.M. Shakesheff, M. Modo, The support of neural stem cells transplanted into stroke-induced brain cavities by PLGA particles, Biomaterials. 30 (2009) 2985-2994.
https://doi.org/10.1016/j.biomaterials.2009.02.012

[112] K. Duncan, G.S. Gonzales-Portillo, S.A. Acosta, Y. Kaneko, C.V. Borlongan, N. Tajiri, Stem cell-paved biobridges facilitate stem transplant and host brain cell interactions for stroke therapy, Brain Res. 1623 (2015) 160-165.
https://doi.org/10.1016/j.brainres.2015.03.007

[113] H. Yu, B. Cao, M. Feng, Q. Zhou, X. Sun, S. Wu, S. Jin, H. Liu, J. Lianhong, Combined transplantation of neural stem cells and collagen type I promote functional recovery after cerebral ischemia in rats, Anat Rec (Hoboken). 293 (2010) 911-917. https://doi.org/10.1002/ar.20941

[114] B. Oh, P. George, Conductive polymers to modulate the post-stroke neural environment, Brain Research Bulletin. 148 (2019) 10-17.
https://doi.org/10.1016/j.brainresbull.2019.02.015

[115] R.A. Sobel, The extracellular matrix in multiple sclerosis lesions, J Neuropathol Exp Neurol. 57 (1998) 205-217. https://doi.org/10.1097/00005072-199803000-00001

[116] S.-Z. Guo, X.-J. Ren, B. Wu, T. Jiang, Preparation of the acellular scaffold of the spinal cord and the study of biocompatibility, Spinal Cord. 48 (2010) 576-581. https://doi.org/10.1038/sc.2009.170

[117] P.M. Crapo, C.J. Medberry, J.E. Reing, S. Tottey, Y. van der Merwe, K.E. Jones, S.F. Badylak, Biologic scaffolds composed of central nervous system extracellular matrix, Biomaterials. 33 (2012) 3539-3547. https://doi.org/10.1016/j.biomaterials.2012.01.044

[118] B.G. Ballios, M.J. Cooke, L. Donaldson, B.L.K. Coles, C.M. Morshead, D. van der Kooy, M.S. Shoichet, A Hyaluronan-Based Injectable Hydrogel Improves the Survival and Integration of Stem Cell Progeny following Transplantation, Stem Cell Reports. 4 (2015) 1031-1045. https://doi.org/10.1016/j.stemcr.2015.04.008

[119] R.K. Mittapalli, X. Liu, C.E. Adkins, M.I. Nounou, K.A. Bohn, T.B. Terrell, H.S. Qhattal, W.J. Geldenhuys, D. Palmieri, P.S. Steeg, Q.R. Smith, P.R. Lockman, Paclitaxel-Hyaluronic NanoConjugates Prolong Overall Survival in a Preclinical Brain Metastases of Breast Cancer Model, Mol Cancer Ther. 12 (2013) 2389-2399. https://doi.org/10.1158/1535-7163.MCT-13-0132

[120] J. Lam, W.E. Lowry, S.T. Carmichael, T. Segura, Delivery of iPS-NPCs to the Stroke Cavity within a Hyaluronic Acid Matrix Promotes the Differentiation of Transplanted Cells, Adv Funct Mater. 24 (2014) 7053-7062. https://doi.org/10.1002/adfm.201401483

[121] N.B. Skop, F. Calderon, S.W. Levison, C.D. Gandhi, C.H. Cho, Heparin crosslinked chitosan microspheres for the delivery of neural stem cells and growth factors for central nervous system repair, Acta Biomater. 9 (2013) 6834-6843. https://doi.org/10.1016/j.actbio.2013.02.043

[122] Y. Wu, W. Wei, M. Zhou, Y. Wang, J. Wu, G. Ma, Z. Su, Thermal-sensitive hydrogel as adjuvant-free vaccine delivery system for H5N1 intranasal immunization, Biomaterials. 33 (2012) 2351-2360. https://doi.org/10.1016/j.biomaterials.2011.11.068

[123] A.J.T. Hyatt, D. Wang, J.C. Kwok, J.W. Fawcett, K.R. Martin, Controlled release of chondroitinase ABC from fibrin gel reduces the level of inhibitory glycosaminoglycan chains in lesioned spinal cord, J Control Release. 147 (2010) 24-29. https://doi.org/10.1016/j.jconrel.2010.06.026

[124] T. Nakaji-Hirabayashi, K. Kato, H. Iwata, In vivo study on the survival of neural stem cells transplanted into the rat brain with a collagen hydrogel that incorporates laminin-derived polypeptides, Bioconjug Chem. 24 (2013) 1798-1804. https://doi.org/10.1021/bc400005m

[125] V.L. Cross, Y. Zheng, N. Won Choi, S.S. Verbridge, B.A. Sutermaster, L.J. Bonassar, C. Fischbach, A.D. Stroock, Dense type I collagen matrices that support cellular remodeling and microfabrication for studies of tumor angiogenesis and vasculogenesis in vitro, Biomaterials. 31 (2010) 8596-8607. https://doi.org/10.1016/j.biomaterials.2010.07.072

[126] D. Jiang, J. Liang, P.W. Noble, Hyaluronan in tissue injury and repair, Annu Rev Cell Dev Biol. 23 (2007) 435-461. https://doi.org/10.1146/annurev.cellbio.23.090506.123337

[127] J. Zhong, A. Chan, L. Morad, H.I. Kornblum, G. Fan, S.T. Carmichael, Hydrogel matrix to support stem cell survival after brain transplantation in stroke, Neurorehabil Neural Repair. 24 (2010) 636-644. https://doi.org/10.1177/1545968310361958

[128] J.S. Lewis, K. Roy, B.G. Keselowsky, Materials that harness and modulate the immune system, MRS Bull. 39 (2014) 25-34. https://doi.org/10.1557/mrs.2013.310

[129] J.E. Rinholm, N.B. Hamilton, N. Kessaris, W.D. Richardson, L.H. Bergersen, D. Attwell, Regulation of oligodendrocyte development and myelination by glucose and lactate, J Neurosci. 31 (2011) 538-548. https://doi.org/10.1523/JNEUROSCI.3516-10.2011

[130] H.K. Makadia, S.J. Siegel, Poly Lactic-co-Glycolic Acid (PLGA) as Biodegradable Controlled Drug Delivery Carrier, Polymers (Basel). 3 (2011) 1377-1397. https://doi.org/10.3390/polym3031377

[131] M.J. Cooke, Y. Wang, C.M. Morshead, M.S. Shoichet, Controlled epi-cortical delivery of epidermal growth factor for the stimulation of endogenous neural stem cell proliferation in stroke-injured brain, Biomaterials. 32 (2011) 5688-5697. https://doi.org/10.1016/j.biomaterials.2011.04.032

Applications of Polymers in Surgery
Materials Research Foundations **123** (2022) 123-154

Materials Research Forum LLC
https://doi.org/10.21741/9781644901892-5

Chapter 5

Applications of Polymers in Ophthalmology

Akhilesh Kumar Tewari[1], Shashi Kiran Misra[2], Kamla Pathak[3]*

[1]Formulation Research and Development, Mankind Research Centre, Manesar, Gurugram, Haryana 122050, India

[2]University Institute of Pharmacy, Chhatrapati Shahu Ji Maharaj University, Kanpur, India, 208026

[3]Faculty of Pharmacy, Uttar Pradesh University of Medical Sciences, Saifai, Etawah 206130, Uttar Pradesh, India

* kamlapathak5@gmail.com

Abstract

Ophthalmology focuses on health issues related to eye, vision, anatomy and concerned ailments. Hitherto, a variety of materials such as ceramic, metal, glass and polymers have been extensively explored that modify optic adherence, residing time, targeted release with minimum downsides. Modern ophthalmic implants including contact lens, ocular endotamponades, viscoelastic replacements etc. are developed by polymers owing to their inherent virtues such as biocompatible with eye tissues, ease of manufacturing, flexible in operation, cost effective etc. The chapter emphasizes on biodegradable (chitosan, poly-lactic acid) and non- biodegradable (PVA, silicones) polymeric ocular preparations such as artificial cornea, orbital ball, glaucoma drainage device etc.

Keywords

Ophthalmology, Polymers, Contact Lens, Glaucoma Drainage, Artificial Cornea

Contents

1.   **Introduction**

Ophthalmology (*ophthalmo=eye*; *logy=study of*) is a branch of medical science concerned with the eyes and their disorders. It deals with the anatomy, physiology and diseases of the eye [1]. The eye and its surrounding structure can be affected by a number of medical conditions such as retinal detachment, diabetic retinopathy, proliferative vitreoretinopathy, cataract, glaucoma etc. [2,3]. Ophthalmology involves diagnosis and treatment of such clinical conditions along with surgery. A wide variety of materials such as metals, ceramics, and polymers have been proposed and used in various clinical ophthalmological conditions. Among the numerous materials, polymers have been extensively explored in ophthalmological applications. Polymer represents several repeating units leading to formation of a compound that have characteristics of high relative molecular mass and associated properties. In 1833 Jons Jakob Berzelius coined term *"polymer"* which originated from Greek word (*polus*, meaning *"many, much"*) and (*meros*, meaning *"part"*).

1.1 Significance of polymers in ophthalmology

Following major developments in polymer fabrication, the molecular properties of polymers were only recognized after the efforts of Hermann Staudinger in 1922 who first suggested that polymers consist of long chains of atoms bound together by covalent bonds. This took more than a decade to achieve acceptance among the scientific community. Initially, polymers were investigated according to the theory of associations or composite theory that emerged in 1861. It was suggested that cellulose and other polymers were colloids and molecular aggregates with small molecular weights bound by an unidentified intermolecular force. Polymers being organic in nature offer a wide range of physical, chemical and mechanical properties which is unmatched by metals and ceramics [4]. Their versatile nature and biocompatibility encouraged extensive research, development and applications in clinical ophthalmology (Fig.1). The use of polymers is not new in the medical field. Most of the modern medical devices are made up of polymeric biomaterials. Their applications in clinical ophthalmology includes contact lenses, intraocular lenses, ocular endotamponades, vitreous replacement fluids, artificial corneas, artificial orbital walls, artificial lacrimal ducts, glaucoma drainage implants, viscoelastic replacements, retinal tacks and adhesives, and scleral buckles [5,6].

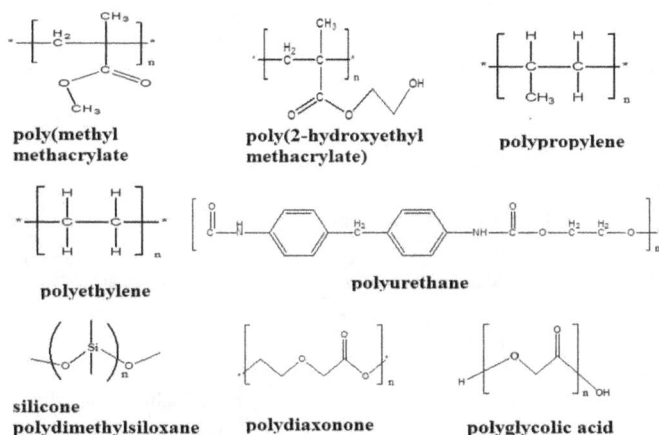

Fig 1. Major polymers explored in development of various ophthalmic devices

1.2   Polymers used in ophthalmogy

Both natural as well as synthetic polymers are used in ophthalmological applications. However, in the last few years special attention has been given to natural polymeric materials with intent to improve biocompatibility which is defined as physical, chemical and biological compatibility of a biomaterial with body tissues leading to no rejection or foreign body reaction. Since 1949, polymethyl methacrylate (PMMA) has been used as standard intraocular lens (IOL) material to treat cataract and other vision deficiencies such as myopia, presbyopia, and hyperopia [7]. Table 1 compiles widely utilized polymers loaded with therapeutic agent for fabrication of ocular implants and devices.

Table 1: Major polymers and their functions in ophthalmology

Polymers	Therapeutic agent	Function	Ref.
Chitosan	Indomethacin, Cyclosporine A	Absorption enhancer	[8]
PLA-PGA	Vancomycin	Biodegradation	[9]
Eudragit	Flurbiprofen	Sustained release	[10]
Alginate	Gatifloxacin	Gelling agent	[11]
Carboxy methyl cellulose	Tropicamide	Viscosity modifier	[12]
Carbopol	Acetazolamide	Improve solubility and ocular absorption	[13]
Silicone	Ciprofloxacin	Controlled release	[14]
Polyvinyl alcohol	Ciprofloxacin	Prolong drug release	[15]
Poly-d-lactic acid/ Polyethylene glycol	Acyclovir	Increase ocular adhesion	[16]
Poly methacrylate	Ciprofloxacin	Prolong drug release	[15]
Poloxamer analogs/carbopol	Puerarin	Improved bioavailability	[17]
Polystyrene	Luteinizing hormone releasing hormone agonist	Targeted sustained release	[18]
Alginate and collagen	Bovine serum albumin	Combat corneal disease	[19]
Hydroxyl propyl methyl cellulose	Gatifloxacin	Viscosity enhancer	[20]
Pluronic F127	Timolol maleate	Enhance bioavailability	[21]
Poly methylidine malonate	Triamcinolone acetonide	Biocompatible ocular implant	[22]
Chitosan/ Polyethylene oxide	Ofloxacin	Enhance transocular permeation	[23]

2. Polymers in contact lens

A contact lens is a prescription medical device which is placed on the front surface of the eye i.e. cornea. They are used as alternate to glasses or spectacles and aid in correcting vision deficiencies not limiting to myopia, hyperopia, astigmatism, and presbyopia [24]. Contact lenses have also been explored medically as therapeutic devices for ophthalmic drug delivery. They have been used to treat certain eye diseases, corneal abrasions, corneal perforations, pain relief in bullous keratopathy and corneal wound healing [25, 26]. There has been a continuous progress in the field of contact lens materials. They demand characteristic physicochemical properties including transparency, flexibility, low density, toughness, physiological inactivity, suitable refractive index, and oxygen permeability. Contact lenses evolved significantly from polymers [27]. The first contact lens was made from PMMA which is a hard, tough and lightweight polymer [28]. It is optically transparent and offers greater clarity than glass. However, contact lenses developed from PMMA suffered poor oxygen permeability. The cornea being an avascular tissue obtains its oxygen requirement through diffusion from atmospheric air (Fig. 2).

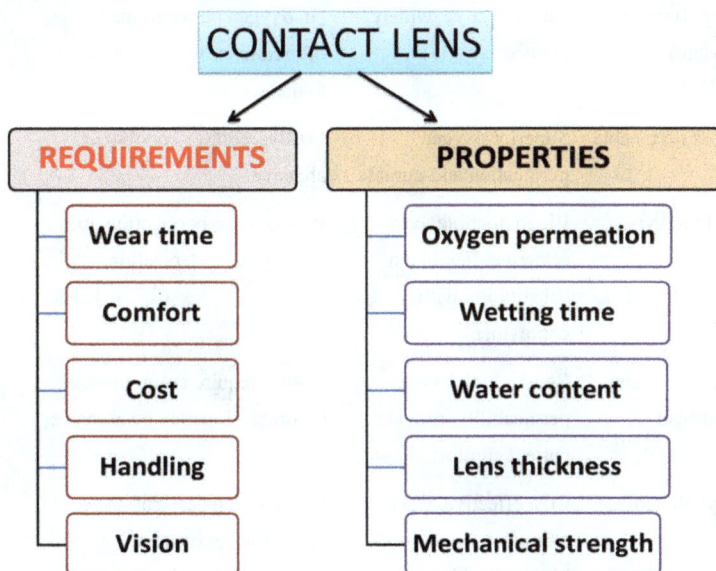

Fig. 2: Requirements and properties of contact lens

Applications of Polymers in Surgery   Materials Research Forum LLC
Materials Research Foundations **123** (2022) 123-154   https://doi.org/10.21741/9781644901892-5

PMMA being rigid in nature acts like a plastic shield and do not allow oxygen to pass through to the cornea resulting in adverse clinical events namely hypoxia and neovascularization. This led to the discovery of a polymer hydrogel made from hydroxyethyl methacrylate. Poly (2-hydroxyethyl methacrylate) synthesized by solution polymerization of 2-hydroxyethyl methacrylate monomer units, is a polymer which forms hydrogel in water. It consists of hydroxyethyl side chain which is hydrophilic and absorbs water due to their electrochemical polarity. The oxygen permeates through the aqueous portion of the hydrogel to fulfill the oxygen demand of the cornea and render them healthy [29]. Researchers also quickly developed polymers with inherent better oxygen permeability. This was achieved by introduction of a siloxane group into the polymer rendering them highly permeable to gases [30]. Table 2 represents the merits and demerits of polymers utilized in fabrication of contact lens.

Table 2. Merits and demerits of polymers used in development of contact lens

Polymers	Merit	Demerit	Ref.
PMMA (poly-meth-methacrylate)	Cost effective, widely applied	No oxygen permeation, inflexible, provide discomfort to the eye	[31]
Rigid glass permeable (RGP)	Superior oxygen permeation and durable	Costly, requires copolymer, abrasive	[32]
Hydroxyl methyl-methacrylate (HEMA) hydrogel	Biocompatible, cost effective, flexible in substitution with copolymer	Poor oxygen permeation, may cause protein deposition	[33]
Silicon and siloxane based hydrogel	Render high oxygen permeability, provide patient comfort, durable	Costly, require co-monomers (hydrophilic), may be abrasive	[34]
Poly vinyl alcohol (PVA)	Cost effective, flexible manufacturing, biocompatible	Poor oxygen permeation, limited water content	[35]

3. Polymers in artificial corneas

The dome-shaped, clear frontal part of our eye which helps to focus light inside of it is referred to as cornea [36]. This avascular transparent connective tissue acts as the primary structural and infectious barrier to the eye. It is made up of several layers including the five recognized layers, three cellular layers namely epithelium, stroma, endothelium and the two interface layer namely Bowman membrane, Descement membrane [37]. Its importance lies in the fact that the cornea provides almost two third refractive power of the eye [38]. According to the World Health Organization, no less than 2.2 billion people around the globe suffer from distant or near vision impairment and uncorrected refractive errors, cataract tops the charts [39]. Corneal diseases can be cured with a prescription pill or an eye drop but once the condition becomes severe, some other treatment is sought for [40]. When the damaged cornea cannot be cured and repaired, the damaged part can be replaced or transplanted with a corneal tissue from a donor. This is done by either penetrating keratoplasty where a full corneal transplant takes place or lamellar keratoplast where only a part of cornea is replaced [41]. In some cases, like patients with chemical burns, Stevens-Johnson's syndrome, limbal stem cell failure and multiple graft failure where the patient can't tolerate the transplanted human corneal tissue or the human donor is not readily available, the option for artificial cornea also exists [42, 43]. Figure 3 displays strategies applied for damage control of cornea.

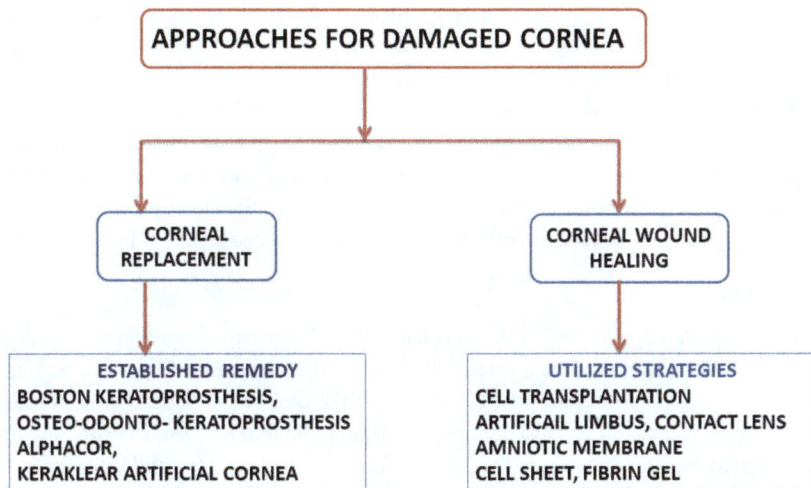

Fig. 3: Promising approaches utilized for damaged cornea

Applications of Polymers in Surgery   Materials Research Forum LLC
Materials Research Foundations **123** (2022) 123-154   https://doi.org/10.21741/9781644901892-5

3.1   KPro artificial cornea

This surgical procedure where the damaged cornea is replaced by an artificial cornea is known as keratoprosthesis or KPro [44]. The concept of artificial cornea was brought in by Guillaume Pellier de Quengsy in the year 1789 [45]. Further down the line, in the year 1856, the first artificial cornea constructed out of glass was implanted into a patient by Nussbaum [46] and as of now multiple keratoprosthesis have been developed.

3.2   B-KPro

The second most clinically efficient keratoprosthesis are Boston Keratoprosthesis also known as B-KPro (Fig. 4). Other keratoprosthesis available are MICOF (Moscow eye microsurgery complex in Russia), Alphacor, Miro Cornea UR, KeraKlear etc. The Boston KPro is available in 2 types. Type I, the most common and effective out of the two KPros. It was developed in the year 1960s at the Massachuesetts Eye and Ear Infirmary and was then approved by FDA in the year 1992. It received the CE-mark in the year 2014 and is till date used as a reference for other KPros [47]. It is a collar button design that serves as an alternative to risky keratoplasty and is constructed in two parts, the anterior plate is made up of the PMMA whereas the back plate is made up of titanium consisting of 16 holes which brings the aqueous humor in contact with corneal tissue. The donor cornea is then exerted in between the 2 plates which is then inserted in the patient's eye. PMMA is a transparent, biologically inert material with high mechanical strength. Its remarkable properties and easy processing led to its extensive use in keratoprosthesis [48, 49]. The type II and the less common, newly introduced device are used for last stage dry eye conditions and ocular diseases such as cicatricial pemiphigoids. Here, the optic is implanted via the closed eye lid and differs from type I by just having a 2 mm anterior nun designed solely to penetrate the lids in between or the skin. The material of construct also remained the same and it was introduced in order to replace the already established OOKP however the visual and retention results were not a huge success [50, 51].

3.3   OOKP

Osteo-Odonto-Keratoprosthesis also known as OOKP keratoprosthesis which has stood the time and is still being used as a reference is the OOKP, introduced by Strampelli in the year 1963 (Fig. 4). This enduring device utilizes the oral mucosa membrane in order to protect the device and an unusual approach of using tooth as a KPro skirt. PMMA cylinder (appropriate power) is inserted in the lamina which is then cemented and is used as a filter rather than an adhesive [52]. A modified version of the OOKP known as modified OOKP (MOOKP) with various improvements was also introduced by Faclinelli at later stages [53]. Apart from its biocompatible properties, the optical cylinder possesses characteristic's like mean extraocular diameter, 3.65 mm (range, 3.3- 4.0 mm); mean radius of the convex intraocular surface, 6.5 mm; mean radius of the convex extraocular surface, 16 mm; equivalent power, 50.8 diopters; mean intraocular diameter, 4.1 mm (range, 3.6-4.6 mm) and refractive index, 1.49. Also, these cylinders provide with an option to increase the theoretical visual field (maximum) [54].

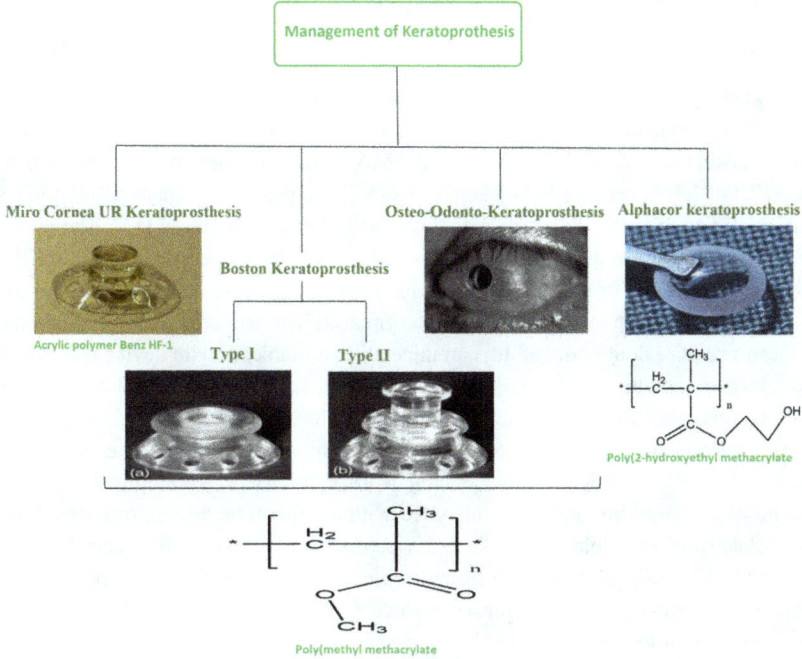

Fig. 4: Various strategies for clinical management of keratoprosthesis

The Boston KPro and OOKP fall in the category of hard keratoprosthesis and a new category of soft keratoprosthesis was introduced with the Alphacor keratoprosthesis. Both the skirt and optic are fabricated from the polymer poly (2-hydroxyethyl methacrylate) also known as PHEMA. The core is composed of the transparent and homogeneous PHEMA whereas the skirt is made up of opaque, porous PHEMA allowing the biointegration with the nearby corneal tissue due to the presence of collagen deposition and cellular colonization. The two regions are joined by the interpenetrating of polymers at the junctional zone referred to as the interpenetrating polymer network. It is currently available in two powers AlphaCor-P for the phakic or the pseudophacic patients and AlphaCor-A for the aphakic patients. PHEMA works by providing the correct refractive power [54]. Its high flexibility, non-toxicity, increased mechanical strength, biocompatibility and water content makes it the first hydrogel in any biomedical devices. Also, PHEMA helps in the existence of the skirt and core keratoprosthesis [49].

3.4   MICOF

MICOF keratoprosthesis developed by the Moscow Eye Microsurgery Complex in Russia involves 2 surgical procedures. First, by inserting a support titanium frame inside the lamellar pocket and second, via implanting PMMA optical cylinder after 3 months (of stage one). MICOF keratoprosthesis is been suggested for post inflammatory, bilateral and posttraumatic conditions which are associated with corneal blindness [55]. The complex surgical procedures (3 stage procedure for OOKP and 2 stage procedure for Boston KPros), the requirement of donor cornea (Boston KPro) and the adverse complications led to the introduction of a non-penetrating artificial cornea known as KeraKlear keratoprosthesis that doesn't require donor cornea. It is an injectable, foldable, acrylic device which comes in a single-piece and lacks any locking ring and back plates. It is constructed from a flexible and biocompatible, strong proprietary intraocular lens material and gives the added advantage of one stage procedure leading to less clinic visits [56]. Another attempt was made to replace the OOKP by introducing Miro Cornea UR Keratoprosthesis for the treatment of coronary blindness accompanied with autoimmune disease or burns. Similar to KeraKlear, it is a flexible, single piece device made up of hydrophobic acrylic polymer developed by CORNEA Consortium from the year 2005 to 2008. Soft hydrophobic acrylic polymer Benz HF-1 was used for constructing this device as it offers mechanical processing even at low temperatures by cryo-milling and cryo-lathing. Also, it possesses mechanical properties that are similar to corneal tissue at 35°C temperature and an added advantage of reducing the mechanical shear between the surrounding tissue and keratoprosthesis haptic during eye rubbing or blinking. Additionally, intra ocular pressure can be easily measured through the haptic using this polymer and wettability of the surface in contact leads to suppressed epithelial cell growth [57].

3.5   Other polymeric keratoprosthesis

Some other keratoprosthesis at different times have also been in the market such as Seoul-type keratoprosthesis (central optic –PMMA, polymer skirt – polypropylene or polyurethane, haptic- monofilament polypropylene), BIOKOP keratoprosthesis (haptic-microporous fluorocarbon, central optic – PMMA), Supra Descemetic keratoprosthesis (hydrophilic hydroxyethyl methacrylate-methyl methacrylate showed the best results

Applications of Polymers in Surgery Materials Research Forum LLC
Materials Research Foundations **123** (2022) 123-154 https://doi.org/10.21741/9781644901892-5

among PMMA, hydroxyethyl methacrylate-N-vinyl pyrrolidone and hydrophilic hydroxyethyl methacrylate-methyl methacrylate; more research needed) [58].

Throughout the history various polymers such as celluloid, PMMA, PHEMA, polytetrafluoroethylene, poly(ethylene tetraphthalate) viz Dacron, polysiloxanes, polyurethanes (from polyols and hexamethylenediisocynate; Quadrol; ethylene oxide, poly (tetramethylene ether)glycol, 4,4'-dicyclohexylmethane diisocyanate) , insoluble poly(vinyl alcohol), poly(glyceryl monomethacrylate) hydrogels, duoptix (a mechanically enhanced hydrogel) etc. [59] have been used with the basic aim to develop a biocompatible, non-toxic or allergic replacement to the natural cornea. Other than the approach to utilize the artificial polymers, natural polymers, implants and bio-engineered cornea can also be used in a regenerative manner to replace the damaged cornea. Some examples are fibrin-based corneal implants, recombinant human collagen corneal implants, silk-based corneal implants, self-assembled corneal implants, grafting of corneal limbus pieces, 2-methacryloxyethyl phosphorylcholine [60, 61]. With the advent of new technology, new alternatives to artificial polymers keep on arising. In a study it was found that the combination of PMMA – PHEMA can serve as a better alternative to them being used individually, novel interpenetrating network based on poly (acrylic acid) and poly (ethylene glycol) gave good results [62], PHEMA incorporated with acrylate comonomer 2-(dimethylamino) ethyl methacrylate (positively charged) and acrylate co-monomer phenoxyethyl methacrylate (hydrophobic) were seen to enhance the innate properties of PHEMA by cell spreading and adhesion [63]. Table 3 summarizes different natural and synthetic polymers employed for the corneal regeneration. Today, we have multiple options varying on severity of condition and the need of treatment. From artificial polymers to regenerative grafts we have come a long way to cure the prevalent and severe corneal blindness and still have a long way to go.

4. Polymers in glaucoma drainage devices

Glaucoma is a disease that damages the optic nerve of the eye leading to loss of vision and blindness. Eye constantly produces aqueous humor, transparent water like fluid which nourishes the lens and maintains the intraocular pressure. As new aqueous humor produces, the same amount drains out through drainage angle to stabilize the pressure in the eye. However, if drainage angle does not function appropriately, the fluid builds up leading to increased intraocular pressure. The rise in pressure inside the eye damages the optic nerve leading to progressive blindness [71]. There are several ways to treat glaucoma. The increased intraocular pressure can be reduced either pharmacologically by use of eye drops or surgically by implanting a glaucoma drainage device.

Applications of Polymers in Surgery   Materials Research Forum LLC
Materials Research Foundations **123** (2022) 123-154   https://doi.org/10.21741/9781644901892-5

Table 3: Polymers employed for corneal regeneration

Polymers	Approach	Outcomes	Ref.
Gelatin/ascorbic acid cryogel	Antioxidant mediated corneal stroma tissue engineering	Biocompatible and improved tissue matrix	[64]
Silk membrane (Aligned)	Neuropeptide embedded multi-layered silk membrane for corneal stroma regeneration	Differentiating periodontal ligament stem cells toward keratocytes	[65]
Chitosan-gelatin hydrogel (thermosensitive)	SDF-1(stromal cell derived factor) embedded corneal epithelial tissue regeneration	Embedded SDF-encouraged reconstruction of corneal tissue	[66]
Silk fibroin	Designing of artificial corneal endothelial tissue	Biocompatible and high mechanical strength corneal tissue	[67]
Peptides (short collagen) linked polyethylene glycol	Development of customized analogs corneal endothelial regeneration	The system stimulated promotion of corneal regeneration via extracellular vesicle production	[68]
Nanofibers containing poly (glycerol sebacate) and poly (ε-caprolactone)	Biodegradable, elastic, corneal endothelial scaffolds	Blended polymeric, elastic, semi-transparent nanofibrous artificial corneal endothelial tissue	[69]
Crosslinked gelatin methacrylate	Nanodimensional, hydrogel pattern corneal endothelial tissue	Transparent and improved mechanical strength, that imparts nutrient transport	[70]

4.1   General description on glaucoma drainage implant

A glaucoma drainage implant is a small medical device which consists of a plate and a tube. The tube is inserted into the eye usually between the cornea and iris. It acts like an artificial drain to let the fluid drain out into the plate. The plate is inserted on the surface of the eye usually underneath the upper eyelid and acts like a pool or reservoir for the fluid. The fluid then slowly percolates through the plate and gets absorbed into the body fluids [72]. Although glaucoma drainage implant is an appropriate therapy to reduce elevated intraocular pressure yet it suffers fibrotic tissue response to the implant material resulting in blockage of the drainage system. The material of construction of the implant plays an

Materials Research Forum LLC
https://doi.org/10.21741/9781644901892-5

essential role to avoid detrimental reactions with living ocular tissues. The materials with suboptimal compatibility results in complications such as fibrinous pupillary membrane occlusion, corneal endothelial cell damage, implant tube migration, extrusion and fibrous encapsulation [73]. Polymers remain the most commonly used material for the glaucoma drainage devices since the introduction of the first polymeric implant, Molteno, in 1979. The endplate of the implant was made of polypropylene and the tube was made of silicone. In 1990, two implants Baerveldt and Krupin immerged which were made of silicone. Ahmed implant emerged in 1993 and the plate of the implant was made of polypropylene with silicone valve. In 1995, Optimed implant was introduced. The endplate of the implant was made of silicone with PMMA matrix (Fig. 5). Ayyala et al studied the inflammatory response associated with endplates made of three different biomaterials namely silicone, polypropylene and a new biocompatible polymer vivathane [74].

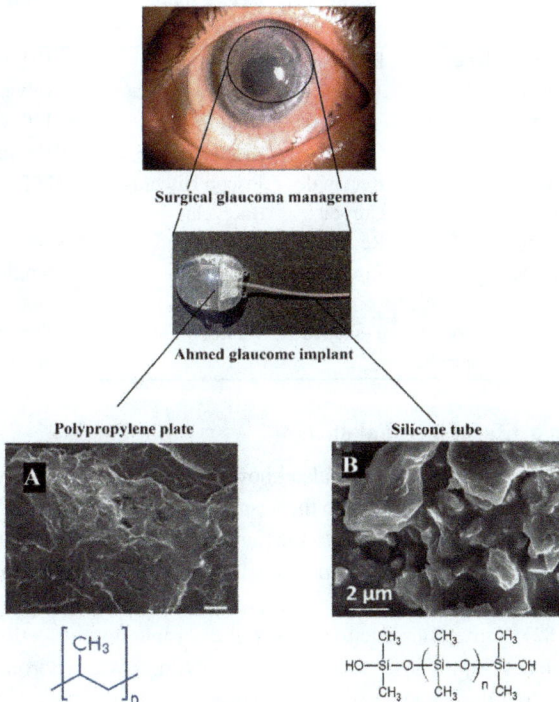

Fig.5: Management of glaucoma through Ahmed glaucoma implant

Elastomeric silicone-(polydimethylsiloxane) has been extensively used in current implants. Seughwan et al. developed a novel glaucoma drainage device for effective regulation of intraocular pressure. The device is made of three biocompatible polymeric layers. All the three layers were made of elastomeric silicone (polydimethylsiloxane) which were bonded together to form a thin implant [75]. Similarly, Chang et al fabricated a polymeric micro check valve for regulation of intraocular pressure. The micro check valve is made of biocompatible polydimethylsiloxane polymer [76]. Other recommended glaucoma regulating strategies are also in market i.e. drug coated glaucoma drainage device [77] and microshunt glaucoma drainage [78, 79]. Table 4 compiled currently used polymeric glaucoma drainage devices with different shape and dimensions.

Table 4. Currently used glaucoma drainage devices and their dimension (Retrieved from Sharaawy et al. 2011) [80]

Type of device	Polymer	Shape	Device name	Dimension (plate area)
Valved	Polypropylene	Pear	Ahmed	184 mm^2 (FP7) venturi valve
	Polymethacrylate	Rectangle	Optimed	140 mm^2 microtubule
	Sialistic	Oval	Krupin disc	180 mm^2 slit valve
	Silicone	Rectangle	Joseph Hitchings	765 mm^2
Non-valved	Silicone	Curved	Baerveldt	250 - 500 mm^2
	Silicone	Rectangle	Shoket band (360°)	300 mm^2
	Silicone	Round	Molteno	Single plate (134 mm^2) Double plate (268 mm^2)
Shunt	Poly-MEGMA hydrophilic acrylic	T-shaped	T-Flux implant	For non-penetrating glaucoma surgery

5. Polymers in artificial lacrimal ducts

The nasolacrimal duct or lacrimal ducts also known as the tear ducts function by carrying tears from the lacrimal sac of our eye into the nasal cavity in our nose. The little tube in the eye through which the tears drain down to the puncta is called the nasolacrimal duct. This duct is part of the lacrimal duct system and the tear enters at the lacrimal punctae and then follows through the canaliculi present in the eyelids. The canaliculi then drain into the lacrimal sac [81, 82]. Sometimes due to reasons like chronic nose infections (like chronic sinusitis), unusual development of face and skull (as in down syndrome), presence of tumors, nose traumas and polyps (like broken nose, tissue scar), conjunctivitis(inflammation and infection of the conjunctiva) and various age related changes this tear duct gets completely or partially obstructed. It is rarely seen that long-term use of eye drops (glaucoma) or the cancer treatments (side effect of chemo and radiation therapy) can also lead to a blocked tear duct [83]. The most common is the congenital blockage that takes place in almost 20% of new born [84]. Tear drainage system is not fully developed and a possibility of duct abnormality exists in infants which can often be cured by simple massage or minor surgeries. The blocked tear duct can easily be identified by excessive tearing of the eye, blurred vision, recurrent redness and inflammation, crusting eyelids, pus or mucus discharge from the surface and lids, painful swelling inside the corner in the eye (Fig. 6). An immediate cure is needed to relieve the irritated and watery eye and to avoid recurrent eye inflammations and infections. As the tears remain in the drainage system, it leads to the growth of fungi, virus and bacteria and promote severe infections like dacryocystitis in any part of the duct [85].

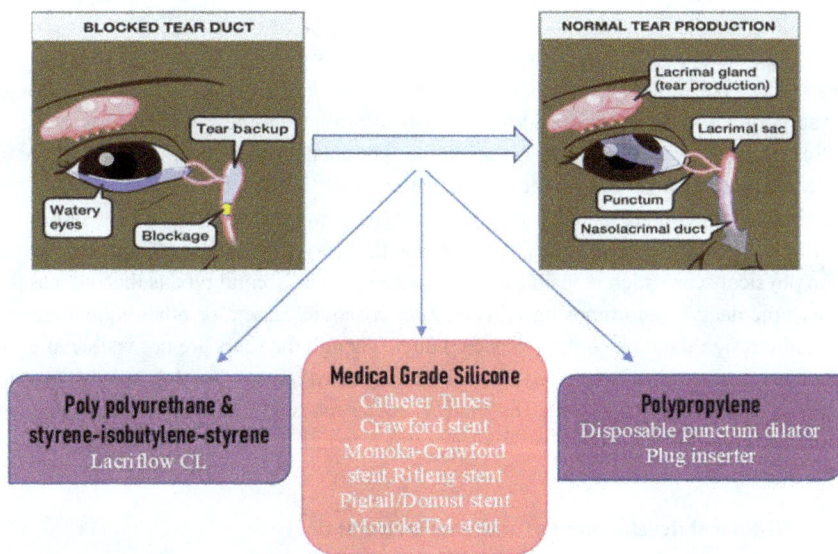

Fig. 6: Polymer based potential strategies utilized for normal tear production

5.1   Strategies to combat blocked duct

Its treatment ranges from placing a wet, warm, clean cloth over the eye to dacryocystorhinostomy, a type of surgery to let the stagnant fluid inside the duct drain out and the easiest way to open a blocked tear duct is by gently massaging it. In case of infants if the blocked tear duct does not open on its own then pressure can be applied through massaging and it can pop open the duct by removing the membrane that originally covered it. In some cases where massaging doesn't cure the problem, doctors can use a thin probe to open up the obstruction in the eye. In severe cases, when probing doesn't prove to be of any success, balloon catheter dilation can be tried. Here, a thin tube known as catheter (mostly made of silicone) is inserted in the tear duct and a balloon is inflated at the other end of the tube. A continuous process of inflation and deflation takes place to widen the duct and release the blockage [86]. Intubation is another process that can help relieve the problem of blocked ducts. A thin silicone tube is taken which is then passed through the opening of the tear duct in the lower and upper eyelid. This tube is inserted to help open the drainage pathway and is kept in place for 3-6 months [87] Intubation is done in cases where the tear duct is partially blocked or failed attempts at probing have already taken place with the symptoms still in place or if a person wants to avoid the surgery or maybe has undergone dacryocystorhinostomy surgery and even after that the duct has blocked again [88]. Dacryocystorhinostomy is a type of surgery to bypass the blocked tear ducts. This is often performed when other treatments fail to yield any positive results. Through, this surgery a new route is created for the tears to drain from the eye. The surgery can be performed in two ways. The first one being the external dacryocystorhinostomy, where the surgeon marks a small cut on the side of nose which is then closed with stitches after inserting a small diameter tube to create a bypass for the tears to flow without any obstruction. 90-95 % success rate is observed with this type of surgery. The one drawback is the physical scar which is visible after the surgery. The second type is the endonasal or endoscopic dacryocystorhinostomy. Here, a microscopic camera or other equipments are inserted through the nose to the obstructed duct. Though the scars are not visible after the surgery and less pain is observed but the success rates are not as high as with external dacryocystorhinostomy [89, 91]. The small diameter tubes which are inserted in the eye to assist in opening the obstructed pathway are also referred as stents and vary based on their place of insertion and usage.

5.2   Historical development of artificial lacrimal duct

Silver wire was used initially by Gaue in the year 1932 as an artificial stent. Then in 1950, Henderson employed polyethylene followed by malleable rods in the year 1962 by Veir [92, 93]. Throughout history, various materials like nylon, silk, dacron, polypropylene have been used in order to better retain and minimize the collateral damage. With the advancing technology, in the year 1968, medical grade silicone was introduced by Keith. Later in 1970 and 1977, Quickert and Dylan and Crawford respectively worked on these tubings and reported bicanalicular stent as a successful attempt [94, 96]. Followed by this, in the year 1989 Bruno fayet and in 1995 Ruban modified the above stents to monocanalicular devices. Then in the year 1998, Rietleng bacanalicular device was further introduced [97,

Applications of Polymers in Surgery
Materials Research Foundations **123** (2022) 123-154

Materials Research Forum LLC
https://doi.org/10.21741/9781644901892-5

98]. Today, we classify the canalicular stents based on different parts of the lacrimal duct system they intubate. The two main classifications of stents as mentioned above are bicanalicular stents which pass through the upper and lower canaliculus forming a closed loop circuit and monocanalicular stents incubate either the lower or upper canaliculus therefore not provide a closed loop system. Numerous stents are available in the market and can be used based on the need and ease of the doctor and patient.

5.3 Stents used to manage blocked lacrimal duct

The material of construct for most of the tubings like Crawford stent, Monoka-Crawford stent, Ritleng stent, Pigtail/Donust stent, MonokaTM stent, O'Donoghue, Guboir and Bika (bicanalicular) stents, Mini-MonokaTM stent, MasterkaTM stent etc. remains to be silicone due to its chemical stability, biocompatibility and dynamic mechanical properties. Medical grade silicone also possesses properties like smooth external surface, simple to use, easily available, pliable, inert, self-retaining, affordable and don't cause any mechanical damage to the nearby eye tissue which make it the ideal material for constructing these stents. Other than this, materials like polyurethane and styrene-isobutylene-styrene (in Lacriflow CL), polypropylene (in disposable punctum dilator and plug inserter), pyrex glass (in Jones tube) etc are also being used for constructing these stents due to their favourable properties like transparency, availability, ease of use, appropriate refractive index, smooth external surface etc. [99, 100]. Table 5 presents generally employed ocular stents and their covering area.

Table 5. General polymeric stent employed for the management of blocked duct (retrieved from Ricca et al. 2018) [101]

Type of stent	Name of stent	Polymeric device	Covering site
Bicanalicular stent	Crawford stent	Silicon rod bearing silk suture	Entire nasolacrimal drainage system
	Ritleng stent	Silicon tube	Entire canalicular system
	Pigtail/donut	Nylon suture	Upper, lower, common canaliculi
Monocanalicular stent	Monoka stent	Silicone	Single canaliculus

Today, multiple treatments are available for blocked lacrimal ducts from the simplest and gentle massage to the complex surgeries employing various artificial polymeric stents. In future many more variations and modifications in the treatment can make this process a lot easier without any pain resulting in almost 100% success rate.

6.   Polymers in artificial orbital walls

The orbit is defined as the skeletal cavity comprising of 7 bones that are situated within the skull. This cavity works by providing mechanical protection to the eye and other soft tissues surrounding it. The orbital walls are the 4 walls that surround the orbit namely superior, medial, inferior and lateral. The superior or the roof wall is the sphenoid bone's lesser wing and frontal bone's orbital part. The medial wall is the ethmoid bone's orbital plate, lacrimal bone, maxilla's frontal process and sphenoid bone's lesser wing. The inferior or the floor wall consists of the orbital surface of zygomatic bone; maxilla and palatine bone [102] whereas the lateral wall contains sphenoid and zygomatic bone [103].

6.1   Kinds of fracture or damage of orbital eye wall

The damage to this wall can be caused when one or more bones surrounding the eyeball break. It can be due to a hard blow on the face causing fracture. Around 85% of eye injuries, including traumatic eye socket fractures are caused by accidents at work place, during sports, in car crashes etc. Around 15% take place due to the violent assaults. Three types of orbital fractures are possible; the first one being the orbital rim fracture caused mainly by car accidents and affects the thick bones of the eye socket present at the outer edge. This is caused by the direct impact on the face and requires a great amount of force to cause injuries to other facial bones and even brain at times [104]. A huge optical impact is seen leading to damage to the optic nerve, eye muscles and nerves. The second can be blowout fracture often caused by getting hit by a fist or with a baseball. The thin inner walls or floor of the orbit breaks apart and the bony structure of the rim of the eye stays intact. Double vision can take place. The third one is the orbital floor fracture and is mainly caused by a fall resulting in the cheek to hit a hard surface. A blow to the eye socket rim can push the bones backward causing the orbital floor bones to buckle downwards. The socket floor and the rim both are fractured [105].

6.2   Symptoms and implants applied for fracture associated with orbital of eye wall

The symptoms vary on the type of fracture and severity of the injury and include facial numbness (due to nerve damage), pain (due to fractures), swelling (result of inflammation), changes in vision (orbital fracture can lead to double vision), bruising (due to the blood pooling), eyeball changes (like sunken eyeball, bloody white part, difficult eye movement). The treatment again depends upon the severity of the damage and it is always advised to let the inflammation decrease before the treatment is started. Ice packs can be used to reduce the inflammation and help in healing the injury by its own [104]. Medications like decongestants and antibiotics can also be prescribed for easing the symptoms and reduce the inflammation and infection. The last resort is to perform a surgery to repair the damages tissue and repair the orbital wall. An implant is used at times to establish a new orbital wall [105]. The goal of an orbital reconstruction surgery is to get back the normal anatomical relations of the orbit while using the perfect implant. The wide variety of grafts available makes this process a little tedious and can be divided into autologous materials, allogenic materials and alloplastic materials [106]. The autologous materials consist of the bone grafts from the donor site and can then be attached with a plate or screw, polyethylene sheet or titanium mesh [107]. This is the most preferred graft due to the strength and rigidity it provides and also minimizes any foreign body reaction. It also possesses certain disadvantages like increased graft reabsorption, donor site morbidity, etc. Other such grafts available are cartilage grafts, periosteal and fascia lata grafts [108]. To bypass the disadvantages of autologous grafts, allogenic materials were introduced as grafts with added advantage of less operating time, lack of donor morbidity, availability of tissue etc. The allogenic materials which have been used over the period are lyophilized cartilage, human dura matter, banked bone, heterogenic bone graft and fascia lata [109-111]. These implants possess a good safety profile; induce osteoconduction, no complications related to the graft etc. Several disadvantages offered by these grafts include higher reabsorption rates than the autologous grafts and possibility of disease transfer like HIV [112-115]. For the reconstruction of the fractured orbit another approach is to use permanent alloplastic materials. They eliminate the donor site morbidity, are readily available and decrease the operative time. Silicone is one such material and is used in form of sheets. Its low cost, ease of handling, flexible nature, tensile strength to provide adequate support are the reasons why it has been used for such a large time. No serious complications have been seen with silicone till date but the risk of poor incorporation of these sheets at the cellular level exists [116-119].

6.3   Polymers/ materials applied for development of implants

Porous polyethylene is another such material which is used as an implant due to its reduced foreign body reactions, tissue and vascular ingrowth and it's easy to work with nature. In some cases, infections followed by removal of the implant were seen with this material [120-122]. Bioactive glass has also been seen as a potential implant and is shown to form a chemical bond with tissues of bone and is osteoconductive. They also showed better outcomes than the cartilage implants in a study and didn't develop any implant related complications and serves as potential material for permanent implant [123-124]. Gelatin film due to its decreased inflammatory response and implant migration, improved healing and no implant-related complications it is also seen as a potential safe and effective implant [125-126]. Polydioxanone, a synthetic biodegradable polymer can be used as an implant and reabsorbs after 6 months following implantation. Biodegradable polyglycolic acid is another option for implant. Titanium is a metallic alloplast and has received great attention due to its malleable and rigid structure [127-130]. The ideal material for an implant is the one that is resistant to infection, reabsorbable, minimally reactive and the most of the materials being used till date have proven to be reliable, safe and effective. Though there is a need in future to develop materials that can overcome the disadvantages being offered by the currently used materials [131].

Conclusion

Eyes, the most sensitive part of the body pose myriad challenges to deliver therapeutic at the site. Emergence of advanced techniques and drug delivery approaches are trying to meet the ophthalmic requirements. Introduction of advent polymeric implants and devices certainly prove potential for zestful release of medicaments having short biological lives, improve therapeutic efficacy and minimize systemic adverse effect for prolong time. Here we discussed versatile polymer based ophthalmic preparations meant for alleviation of common ailments associated with ophthalmic system.

References

[1] J.A. Calles, J. Bermúdez, E. Vallés, D. Allemandi, S. Palma, Polymers in Ophthalmology, in: F. Puoci (Eds.), Advanced Polymers in Medicine, Springer International Publishing, Switzerland, 2015, pp. 147-176. https://doi.org/10.1007/978-3-319-12478-0_6

[2] S. Donati, S.M. Caprani, G. Airaghi, R. Vinciguerra, L. Bartalena, F. Testa, C. Mariotti, G. Porta, F. Simonelli, C. Azzolini, Vitreous substitutes: the present and the future. Biomed Res Int. 2014 (2014) 1-12. https://doi.org/10.1155/2014/351804

[3] Fizia-Orlicz, M. Misiuk-Hojło, The use of polymers for intraocular lenses in cataract surgery (Soczewki wewnątrzgałkowe w chirurgii zaćmy - wykorzystanie polimerów). Polim Med. 45 (2015) 95-102. https://doi.org/10.17219/pim/61978

[4] W. He, R. Benson, 8 - Polymeric Biomaterials, in: M. Kutz (Eds.), Plastics Design Library, Applied Plastics Engineering Handbook (Second Edition), William Andrew Publishing, New York, 2017, pp. 145-164. https://doi.org/10.1016/B978-0-323-39040-8.00008-0

[5] E. Drevon-Gaillot, Ocular medical devices: histologic technique and histopathologic evaluation of the biocompatibility and performance. Toxicol Pathol. 47 (2019) 418-425. https://doi.org/10.1177/0192623318813533

[6] M.F. Maitz, Applications of synthetic polymers in clinical medicine. Biosurface Biotribology. 1 (2015) 161-176. https://doi.org/10.1016/j.bsbt.2015.08.002

[7] P. Özyol, E. Özyol, F. Karel, Biocompatibility of intraocular lenses. Turk J Ophthalmol. 47 (2017) 221-225. https://doi.org/10.4274/tjo.10437

[8] A.A Badawi, H.M El-Laithy, R.K.El Qidra, H. El Mofty, M. El dally, Chitosan based nanocarriers for indomethacin ocular delivery. Arch Pharm Res.8 (2008)1040-9. https://doi.org/10.1007/s12272-001-1266-6

[9] E. Gavini, P.Chetoni, M.Cossu, M.G. Alvarez, M.F. Saettone, P.Giunchedi, PLGA microspheres for the ocular delivery of a peptidedrug, vancomycin using emulsification/spray-drying as the preparationpreparation method: in vitro/in vivo studies. Eur J Pharm Biopharm. 57(2003) 207-12. https://doi.org/10.1016/j.ejpb.2003.10.018

[10] R. Pignatello, C. Bucolo, G. Spedalieri, A. Maltese, G. Puglisi, Flurbiprofen- loaded acrylate polymer nanosuspensions for ophthalmic application. Biomaterials. 23(2002) 3247-55. https://doi.org/10.1016/S0142-9612(02)00080-7

[11] Z. Liu, J. Li, S. Nie, H. Liu, P.Ding, W.Pan, Study of an alginate/HPMC-based in situ gelling ophthalmic delivery system for gatifloxacin. Int J Pharm 2006; 315: 12-7. https://doi.org/10.1016/j.ijpharm.2006.01.029

[12] R. Herrero-Vanrell, A. Fernandez-Carballido, G. Frutos, R.Cadórniga, Enhancement of the mydriatic response to tropicamide by bioadhesive polymers. J Ocul Pharmacol Ther.16 (2000) 419-28. https://doi.org/10.1089/jop.2000.16.419

[13] D.Aggarwal, D. Pal, A.K. Mitra, I.P.Kaur, Study of the extent of ocular absorption of acetazolamide from a developed niosomal formulation, by microdialysis sampling of aqueous humor. Int J Pharm 338 (2007) 21-6. https://doi.org/10.1016/j.ijpharm.2007.01.019

[14] A. Hui, A. Boone, L. Jones, Uptake and release of ciprofloxacin-HCl from conventional and silicone hydrogel contact lens materials. Eye Contact Lens. 34(2008) 266-71. https://doi.org/10.1097/ICL.0b013e3181812ba2

[15] L. Budai, M. Hajdu, M. Budai, P. Grof, S. Beni, B. Noszal, Gels and liposomes in optimized ocular drug delivery: Studies on ciprofloxacin formulations. Int J Pharm 343 (2007) 34-40. https://doi.org/10.1016/j.ijpharm.2007.04.013

[16] C. Giannavola, C. Bucolo, A. Maltese, D. Paolino, M.A.Vandelli, G. Puglisi, Influence of preparation conditions on acyclovirloaded poly-d,l-lactic acid nanospheres and effect of PEG coating on ocular drug bioavailability. Pharm Res 20 (2003) 584-90.

[17] H. Qi, W. Chen, C. Huang, L. Li, C. Chen, W. Li, Development of a poloxamer analogs/carbopol-based in situ gelling and mucoadhesive ophthalmic delivery system for puerarin. Int J Pharm. 337(2007) 178-87. https://doi.org/10.1016/j.ijpharm.2006.12.038

[18] U.B. Kompella, S.S. Raghavan, E.RS. Escobar, LHRH agonist and transferring functionalization enhance nanoparticle delivery in a novel bovine ex vivo eye model. Mol Vis 12 (2006) 1185-98.

[19] W. Liu, M. Griffith, F. Li, Alginate microsphere-collagen composite hydrogel for ocular drug delivery and implantation. J Mater Sci Mater Med. 19(2008) 3365-71. https://doi.org/10.1007/s10856-008-3486-2

[20] Z. Liu, J. Li, S. Nie, H. Liu, P. Ding, W. Pan, Study of an alginate/ HPMC-based in situ gelling ophthalmic delivery system for gatifloxacin. Int J Pharm. 315(2006) 12-7. https://doi.org/10.1016/j.ijpharm.2006.01.029

[21] A.H. El-Kamel, In vitro and in vivo evaluation of Pluronic F127- based ocular delivery system for timolol maleate. Int J Pharm 241(2002) 47-55. https://doi.org/10.1016/S0378-5173(02)00234-X

[22] O. Felt-Baeyens, S. Eperon, P. Mora, D. Limal, S. Sagodira, P. Breton, Biodegradable scleral implants as new triamcinolone acetonide delivery systems. Int J Pharm 322(2006) 6-12. https://doi.org/10.1016/j.ijpharm.2006.05.053

[23] G. Di Colo, Y. Zambito, S. Burgalassi, A. Serafini, M.F.Saettone, Effect of chitosan on in vitro release and ocular delivery of ofloxacin from erodible inserts based on poly(ethylene oxide). Int JPharm 248(2002) 115-22. https://doi.org/10.1016/S0378-5173(02)00421-0

[24] F. Findik, A case study on the selection of materials for eye lenses. Int Sch Res Notices. 2011 (2011) 1-4. https://doi.org/10.5402/2011/160671

[25] M. Rubinstein, Applications of contact lens devices in the management of corneal disease. Eye. 17 (2003) 872-876. https://doi.org/10.1038/sj.eye.6700560

[26] L. Lim, E.W.L. Lim, Therapeutic contact lenses in the treatment of corneal and ocular surface diseases-a review. Asia Pac J Ophthalmol (Phila). 9 (2020) 524-532. https://doi.org/10.1097/APO.0000000000000331

[27] Linkon, Contact lenses and care, Material properties for contact lenses, https://www.likon.com.ua/patients/for-professionals/material-properties-for-contact-lenses/, 2021 (accessed 27 August 2021).

[28] T.B. Harvey, W.B. Meyers, L.M. Bowman, Contact Lens Materials: Their Properties and Chemistries, in: C.G. Gebelein, R.L. Dunn (Eds.), Progress in Biomedical Polymers. Springer, Boston, MA., 1990, pp. 1-2. https://doi.org/10.1007/978-1-4899-0768-4_1

[29] C.S.A. Musgrave, F. Fang, Contact lens materials: a materials science perspective. Materials (Basel). 12 (2019) 1-35. https://doi.org/10.3390/ma12020261

[30] P.C. Nicolson, J. Vogt, Soft contact lens polymers: an evolution. Biomaterials. 22 (2001) 3273-3283. https://doi.org/10.1016/S0142-9612(01)00165-X

[31] A.S. Bruce, N.A. Brennan, R.G. Lindsay, Diagnosis and mangement of ocular changes during contact lens wear, Part II. Clin. Signs Ophthalmol.17(1995) 2-11.

[32] J.L. Alió, J.I. Belda, A. Artola, M. García-Lledó, A. Osman, Contact lens fitting to correct irregular astigmatism after corneal refractive surgery. J Cataract Refract Surg. 28(2002) 1750-7. https://doi.org/10.1016/S0886-3350(02)01489-X

[33] O. Wichterle, D. Lím, Hydrophilic Gels for Biological Use. Nature. 185(1960)117. https://doi.org/10.1038/185117a0

[34] P. Keogh, J.F. Kunzler, G.C.C. Niu, Hydrophilic Contact Lens Made from Polysiloxanes Which Are Thermally Bonded to Polymerizable Groups and Which Contain Hydrophilic Sidechains. 4260725. U.S. Patent. 1981 Apr 7.

[35] M. Kita, Y. Ogura, Y. Honda, S.H. Hyon, W. Cha, Evaluation of polyvinyl alcohol hydrogel as a soft contact lens material. Graefes Arch Clin Exp Ophthalmol. 228(1990)533-7. https://doi.org/10.1007/BF00918486

[36] American Academy of Ophthalmology, Eye Anatomy: Parts of the Eye and How We See. https://www.aao.org/eye-health/anatomy/parts-of-eye, 2021 (accessed 10 August 2021).

[37] D.W. Del Monte, T. Kim, Anatomy and physiology of the cornea. J Cataract Refract Surg. 37 (2011) 588-598. https://doi.org/10.1016/j.jcrs.2010.12.037

[38] H.J. Kaplan, Anatomy and function of the eye. Chem Immunol Allergy. 92 (2007) 4-10. https://doi.org/10.1159/000099236

Materials Research Forum LLC
https://doi.org/10.21741/9781644901892-5

[39] World Health Organization, Blindness and vision impairment. https://www.who.int/en/news-room/fact-sheets/detail/blindness-and-visual-impairment, 2021 (accessed 12 August 2021).

[40] National Eye Institute, Corneal Conditions. https://www.nei.nih.gov/learn-about-eye-health/eye-conditions-and-diseases/corneal-conditions#1, 2021 (accessed 12 August 2021).

[41] National Eye Institute, Corneal Transplants. https://www.nei.nih.gov/learn-about-eye-health/eye-conditions-and-diseases/corneal-conditions/corneal-transplants, 2021 (accessed 12 August 2021).

[42] F.C. Lam, C. Liu, The future of keratoprostheses (artificial corneae). Br J Ophthalmol. 95 (2011) 304-305. https://doi.org/10.1136/bjo.2010.188359

[43] Cornea Research Foundation of America, Artificial Cornea. http://www.cornea.org/Learning-Center/Cornea-Transplants/Artificial-Cornea.aspx, 2021 (accessed 12 August 2021).

[44] University of Iowa Health Care, Keratoprosthesis. https://eyerounds.org/tutorials/Cornea-Transplant-Intro/6-kprosth.htm, 2021 (accessed 13 August 2021).

[45] J.J. Barnham, M.J. Roper-Hall, Keratoprosthesis: a long-term review. Br J Ophthalmol. 67 (1983) 468-474. https://doi.org/10.1136/bjo.67.7.468

[46] M. Bhattacharya, A. Sadeghi, S. Sarkhel, M. Hagström, S. Bahrpeyma, E. Toropainen, S. Auriola, A. Urtti, Release of functional dexamethasone by intracellular enzymes: a modular peptide-based strategy for ocular drug delivery. J Control Release. 327 (2020) 584-594. https://doi.org/10.1016/j.jconrel.2020.09.005

[47] B. Salvador-Culla, P.E. Kolovou, Keratoprosthesis: a review of recent advances in the field. J Funct Biomater. 7 (2016) 1-13. https://doi.org/10.3390/jfb7020013

[48] T.V. Chirila, C.R. Hicks, P.D. Dalton, S. Vijayasekaran, X. Lou, Y. Hong, A.B. Clayton, B.W. Ziegelaar, J.H. Fitton, S. Platten, G.J. Crawford, I.J. Constable, Artificial cornea. Prog Polym Sci. 23 (1998) 447-473. https://doi.org/10.1016/S0079-6700(97)00036-1

[49] Y. Hwang, G. Kim. Evaluation of stability and biocompatibility of PHEMA-PMMA keratoprosthesis by penetrating keratoplasty in rabbits. Lab Anim Res. 32 (2016) 181-186. https://doi.org/10.5625/lar.2016.32.4.181

[50] C.H. Dohlman, M. Harissi-Dagher, B.F. Khan, K. Sippel, J.V. Aquavella, J.M. Graney, Introduction to the use of the Boston keratoprosthesis. Expert Rev Ophthalmol. 1 (2006) 41-48. https://doi.org/10.1586/17469899.1.1.41

[51] A.S. Bardan, N.A. Raqqad, M. Zarei-Ghanavati, C. Liu, The role of keratoprostheses. Eye (Lond). 32 (2018) 7-8. https://doi.org/10.1038/eye.2017.287

[52] M. Zarei-Ghanavati, V. Avadhanam, A.V. Perez, C. Liu, The osteo-odonto-keratoprosthesis. Curr Opin Ophthalmol. 28 (2017) 397-402. https://doi.org/10.1097/ICU.0000000000000388

[53] G.S. Sarode, S.C. Sarode, J.S. Makhasana, Osteo-odonto-keratoplasty: a review. J Clinic Experiment Ophthalmol. 2 (2011) 1-5. https://doi.org/10.4172/2155-9570.1000188

[54] N. Jirásková, P. Rozsival, M. Burova, M. Kalfertova, AlphaCor artificial cornea: clinical outcome. Eye (Lond). 25 (2011) 1138-1146. https://doi.org/10.1038/eye.2011.122

[55] Y. Huang, J. Yu, L. Liu, G. Du, J. Song, H. Guo, Moscow eye microsurgery complex in Russia keratoprosthesis in Beijing. Ophthalmology. 118 (2011) 41-46. https://doi.org/10.1016/j.ophtha.2010.05.019

[56] R. Pineda, Y. Shiuey, The keraklear artificial cornea: a novel keratoprosthesis. Techniques in Ophthalmology. 7 (2009) 101-104. https://doi.org/10.1097/ITO.0b013e3181bee614

[57] G. Duncker, J. Storsberg, W. Müller-Lierheim, The fully synthetic, bio-coated MIRO® CORNEA UR keratoprosthesis: development, preclinical testing, and first clinical results. Spektrum Augenheilkd. 28 (2014) 250-260. https://doi.org/10.1007/s00717-014-0243-4

[58] O. Ilhan-Sarac, E.K. Akpek, Current concepts and techniques in keratoprosthesis. Curr Opin Ophthalmol. 16 (2005) 246-250. https://doi.org/10.1097/01.icu.0000172829.33770.d3

[59] B. Wright, C.J. Connon, Corneal Regenerative Medicine, first ed., Humana Press, Totowa, NJ, 2013. https://doi.org/10.1007/978-1-62703-432-6

[60] R. Singh, N. Gupta, M. Vanathi, R. Tandon, Corneal transplantation in the modern era. Indian J Med Res. 150 (2019) 7-22. https://doi.org/10.4103/ijmr.IJMR_141_19

[61] M. Griffith, E.I. Alarcon, I. Brunette, Regenerative approaches for the cornea. J Intern Med. 280(2016) 276-86. https://doi.org/10.1111/joim.12502

[62] L. Hartmann, K. Watanabe, L.L. Zheng, C.Y. Kim, S.E. Beck, P. Huie, J. Noolandi, J.R. Cochran, C.N. Ta, C.W. Frank, Toward the development of an artificial cornea: improved stability of interpenetrating polymer networks. J Biomed Mater Res B Appl Biomater. 98 (2011) 8-17. https://doi.org/10.1002/jbm.b.31806

[63] S.R. Sandeman, R.G. Faragher, M.C. Allen, C. Liu, A.W. Lloyd, Novel materials to enhance keratoprosthesis integration. Br J Ophthalmol. 84 (2000) 640-644. https://doi.org/10.1136/bjo.84.6.640

[64] L.J. Luo, J.Y. Lai, S.F. Chou, Y.J. Hsueh, D.H. Ma, Development of gelatin/ascorbic acid cryogels for potential use in corneal stromal tissue engineering. Acta Biomater. 65 (2018) 123-136. https://doi.org/10.1016/j.actbio.2017.11.018

[65] J. Chen, W. Zhang, P. Kelk, L.J. Backman, P. Danielson, Substance P and patterned silk biomaterial stimulate periodontal ligament stem cells to form corneal stroma in a bioengineered threedimensional model. Stem Cell Res Ther.260 (2017). https://doi.org/10.1186/s13287-017-0715-y

[66] Q. Tang, C. Luo, B. Lu, Q. Fu, H. Yin, Z. Qin, Thermosensitive chitosan-based hydrogels releasing stromal cell derived factor-1 alpha recruit MSC for corneal epithelium regeneration. Acta Biomater. 61(2017) 101-113. https://doi.org/10.1016/j.actbio.2017.08.001

[67] N. Vázquez, C.A. Rodríguez-Barrientos, S.D. Aznar-Cervantes, M. J. Chacón, L. Cenis A.C. Riestra, Silk fibroin films for corneal endothelial regeneration: transplant in a rabbit descemet membrane endothelial keratoplasty. Invest Ophthalmol Visual Sci. 58 (2017) 3357-3365. https://doi.org/10.1167/iovs.17-21797

[68] J.R. Jangamreddy, M.K.C. Haagdorens, M. Islam, P. Lewis, A. Samanta, P. Fagerholm, Short peptide analogs as alternatives to collagen in pro-regenerative corneal implants. Acta Biomater. 69 (2018) 120-130. https://doi.org/10.1016/j.actbio.2018.01.011

[69] S. Salehi, M. Czugala, P. Stafiej, M. Fathi, T. Bahners, J. S. Gutmann, Poly (glycerol sebacate)-poly (+-caprolactone) blend nanofibrous scaffold as intrinsic bio-and immunocompatible systemfor corneal repair. Acta Biomater. 50 (2017) 370-380. https://doi.org/10.1016/j.actbio.2017.01.013

[70] M. Rizwan, G. S. L. Peh, H.P. Ang, N.C. Lwin, K. Adnan, J.S. Mehta, Sequentially-crosslinked bioactive hydrogels as nanopatterned substrates with customizable stiffness and degradation for corneal tissue engineering applications. Biomaterials 120 (2017) 139-154. https://doi.org/10.1016/j.biomaterials.2016.12.026

[71] American Academy of Ophthalmology, What is Glaucoma? https://www.aao.org/eye-health/diseases/what-is-glaucoma, 2021 (accessed 27 August 2021).

[72] P. Agrawal, P. Bhardwaj, Glaucoma drainage implants. Int J Ophthalmol. 13 (2020) 1318-1328. https://doi.org/10.18240/ijo.2020.08.20

[73] K.S. Lim, B.D.S. Allan, A.W. Lloyd et al. Glaucoma drainage devices; past, present, and future. British Journal of Ophthalmology. 82 (1998) 1083-1089. https://doi.org/10.1136/bjo.82.9.1083

[74] R.S. Ayyala, L.E. Harman, B. Michelini-Norris, et al. Comparison of different biomaterials for glaucoma drainage devices. Arch Ophthalmol. 117 (1999) 233-236. https://doi.org/10.1001/archopht.117.2.233

[75] S. Moon, S. Im, J. An, C.J. Park, H.G. Kim, S.W. Park, H.I. Kim, J.H. Lee, Selectively bonded polymeric glaucoma drainage device for reliable regulation of intraocular pressure. Biomed Microdevices. 14 (2012) 325-335. https://doi.org/10.1007/s10544-011-9609-4

[76]C.J. Park, D.S. Yang, J.J. Cha, J.H. Lee, Polymeric check valve with an elevated pedestal for precise cracking pressure in a glaucoma drainage device. Biomed Microdevices. 18 (2016) 20. https://doi.org/10.1007/s10544-016-0048-0

[77] N. Sahiner, D.J. Kravitz, R. Qadir, D.A. Blake, S. Haque, V.T. John, C.E. Margo, R.S. Ayyala, Creation of a drug-coated glaucoma drainage device using polymer technology: in vitro and in vivo studies. Arch Ophthalmol. 127 (2009) 448-453. https://doi.org/10.1001/archophthalmol.2009.19

[78] C. Wischke, A.T. Neffe, B.D. Hanh, C.F. Kreiner, K. Sternberg, O. Stachs, R.F. Guthoff, A. Lendlein, A multifunctional bilayered microstent as glaucoma drainage device. J Control Release. 172 (2013) 1002-1010. https://doi.org/10.1016/j.jconrel.2013.10.021

[79] M. Löbler, K. Sternberg, O. Stachs, R. Allemann, N. Grabow, A. Roock, C.F. Kreiner, D. Streufert, A.T. Neffe, B.D. Hanh, A. Lendlein, K.P. Schmitz, R. Guthoff, Polymers and drugs suitable for the development of a drug delivery drainage system in glaucoma surgery. J Biomed Mater Res B Appl Biomater. 97 (2011) 388-395. https://doi.org/10.1002/jbm.b.31826

[80] T. Sharaawy, S. Bhartiya, Surgical management of glaucoma: evolving paradigms. Indian J Ophthalmol. 59(2011) 123-30. https://doi.org/10.4103/0301-4738.73692

[81] Cleveland Clinic, Tear System.
https://my.clevelandclinic.org/health/diseases/17540-tear-system, 2021 (accessed 17 August 2021).

[82] F. Paulsen, The human nasolacrimal ducts. Adv Anat Embryol Cell Biol. 170 (2003) 1-106. https://doi.org/10.1007/978-3-642-55643-2_1

[83]American Academy of Ophthalmology, Blocked Tear Duct Causes.
https://www.aao.org/eye-health/diseases/blocked-tear-duct-cause, 2021 (accessed 18 August 2021).

[84]American Academy of Ophthalmology, What Is a Blocked Tear Duct?
https://www.aao.org/eye-health/diseases/what-is-blocked-tear-duct, 2021 (accessed 18 August 2021).

[85] Mayo Clinic, Blocked tear duct, Symptoms and causes. https://www.mayoclinic.org /diseases-conditions/blocked-tear-duct/symptoms-causes/syc-20351369, 2021 (accessed 18 August 2021).

[86] WebMD, Blocked Tear Duct? What to Expect From Treatment.
https://www.webmd.com/eye-health/blocked-tear-ducts, 2021 (accessed 19 August 2021).

[87] UW Health, Blocked Tear Duct and Siicone Intubation.
https://patient.uwhealth.org/healthfacts/7514, 2021 (accessed 19 August 2021).

[88]Dobbs Ferry Pharmacy, Blocked Tear Ducts: Intubation.
http://www.dobbsferrypharmacy.com/PatientPortal/MyPractice.aspx?UAID=C02C914 0-EE45-4012-81AE-26B3E0C445ED&ID=HW5hw208973, 2021 (accessed 19 August 2021).

[89]American Academy of Ophthalmology, EyeWiki, Dacryocystorhinostomy.
https://eyewiki.aao.org/Dacryocystorhinostomy, 2021 (accessed 19 August 2021).

[90] Mayo Clinic, Blocked tear duct, Diagnosis and treatment.
https://www.mayoclinic.org/diseases-conditions/blocked-tear-duct/diagnosis-treatment/drc-20351375, 2021 (accessed 19 August 2021).

[91] J.W. Henderson, Management of strictures of the lacrimal canaliculi with polyethylene tubes. Arch Ophthalmol.44 (1950) 198-203.
https://doi.org/10.1001/archopht.1950.00910020203002

[92] E.R. Veirs, Malleable rods for immediate repair of the traumatically severed lacrimal canaliculus. Trans Am Acad Ophthalmol Otolaryngol. 66 (1962) 263-264.

[93] M.H. Quickert, R.M. Dryden, Probes for intubation in lacrimal drainage. Trans Am Acad Ophthalmol Otolaryngol. 74 (1970) 431-433.

[94] C.G. Keith, Intubation of the lacrimal passages. Am J Ophthalmol. 65 (1968) 70-74. https://doi.org/10.1016/0002-9394(68)91031-3

[95] J.S. Crawford, Intubation of obstructions in the lacrimal system. Can J Ophthalmol. 12 (1977) 289-292.

[96] B. Fayet, J.A. Bernard, Y. Pouliquen, Repair of recent canalicular wounds using a monocanalicular stent (Réparation des plaies canaliculaires récentes avec une sonde mono-canaliculaire à fixation méatique). Bull Soc Ophtalmol Fr. 89 (1989) 819-825.

[97] J.M. Ruban, B. Guigon, V. Boyrivent, Analysis of the efficacy of the large mono-canalicular intubation stent in the treatment of lacrimation caused by congenital obstruction of the lacrimal ducts in infants. J Fr Ophtalmol. 18 (1995) 377-383.

[98] M.R. Pe, J.D. Langford, J.V. Linberg, T.L. Schwartz, N. Sondhi, Ritleng intubation system for treatment of congenital nasolacrimal duct obstruction. Arch Ophthalmol. 116 (1998) 387-391. https://doi.org/10.1001/archopht.116.3.387

[99] Eye Rounds.org, Nasolacrimal Stents: An Introductory Guide. https://webeye.ophth.uiowa.edu/eyeforum/tutorials/Stents/index.htm, 2021 (accessed 19 August 2021).

[100] M. Singh, S. Kamal, V. Sowmya, Lacrimal stents and intubation systems: an insight. DJO. 26 (2015) 14-19. https://doi.org/10.7869/djo.128

[101] A.M. Ricca, A.K. Kim, H.S. Chahal, E.M. Shriver, Nasolacrimal Stents: An Introductory Guide. EyeRounds.org. posted January 29, 2018; Available from: http://EyeRounds.org/tutorials/Stents/

[102] Kenhub, Bones of the orbit. https://www.kenhub.com/en/library/anatomy/bones-of-the-orbit, 2021(accessed 21 August 2021).

[103] C. René, Update on orbital anatomy. Eye. 20 (2006) 1119-1129. https://doi.org/10.1038/sj.eye.6702376

[104] Temple Health, Orbital Fractures. https://www.templehealth.org/services/conditions/orbital-fractures, 2021 (accessed 21 August 2021).

[105] Harvard Health Publishing, Harvard Medical School, Eye Socket Fracture (Fracture Of The Orbit). https://www.health.harvard.edu/a_to_z/eye-socket-fracture-fracture-of-the-orbit-a-to-z, 2021 (accessed 21 August 2021).

[106] R.D. Glassman, P.N. Manson, C.A. Vanderkolk, N.T. Iliff, M.J. Yaremchuk, P. Petty, C.R. Defresne, B.L. Markowitz, Rigid fixation of internal orbital fractures. Plast Reconstr Surg. 86 (1990) 1103-1109. https://doi.org/10.1097/00006534-199012000-00009

[107] K. Hwang, Y. Kita, Alloplastic template fixation of blow-out fracture. J Craniofac Surg. 13 (2002) 510-512. https://doi.org/10.1097/00001665-200207000-00006

[108] K. Chowdhury, G.E. Krause, Selection of materials for orbital floor reconstruction. Arch Otolaryngol Head Neck Surg. 124 (1998) 1398-1401. https://doi.org/10.1001/archotol.124.12.1398

[109] B. Çeliköz, H. Duman, N. Selmanpakoğlu, Reconstruction of the orbital floor with lyophilized tensor fascia lata. J Oral Maxillofac Surg. 55 (1997) 240-244. https://doi.org/10.1016/S0278-2391(97)90533-4

[110]J.M. Neigel, P.O. Ruzicka, Use of demineralized bone implants in orbital and craniofacial reconstruction and a review of the literature. Ophthalmic Plast Reconstr Surg. 12 (1996) 108-120. https://doi.org/10.1097/00002341-199606000-00005

[111]P.D. Waite, J.T. Clanton, Orbital floor reconstruction with lyophilized dura. J Oral Maxillofac Surg. 46 (1988) 727-730. https://doi.org/10.1016/0278-2391(88)90180-2

[112]M.F. Guerra, J.S. Pérez, F.J. Rodriguez-Campo, L.N. Gías, Reconstruction of orbital fractures with dehydrated human dura mater. J Oral Maxillofac Surg. 58 (2000) 1361-1366. https://doi.org/10.1053/joms.2000.18266

[113]J.R. Bevivino, P.N. Nguyen, L.J. Yen, Reconstruction of traumatic orbital floor defects using irradiated cartilage homografts. Ann Plast Surg. 33 (1994) 32-37. https://doi.org/10.1097/00000637-199407000-00007

[114]J.M. Chen, M. Zingg, K. Laedrach, J. Raveh, Early surgical intervention for orbital floor fractures: A clinical evaluation of lyophilized dura and cartilage reconstruction. J Oral Maxillofac Surg. 50 (1992) 935-941. https://doi.org/10.1016/0278-2391(92)90049-6

[115]S. Morax, T. Hurbli, R. Smida, Bovine heterologous bone graft in orbital surgery (Greffe osseuse hétérologue d'origine bovine dans la chirurgie orbitaire). Ann Chir Plast Esthet. 38 (1993) 445-450.

[116]R.H. Haug, E. Nuveen, T. Bredbenner, An evaluation of the support provided by common internal orbital reconstruction materials. J Oral Maxillofac Surg. 57 (1999) 564-570. https://doi.org/10.1016/S0278-2391(99)90076-9

Applications of Polymers in Surgery   Materials Research Forum LLC
Materials Research Foundations **123** (2022) 123-154   https://doi.org/10.21741/9781644901892-5

[117]J.A. Fialkov, C. Holy, C.R. Forrest, J.H. Phillips, O.M. Antonyshyn, Postoperative infections in craniofacial reconstructive procedures. J Craniofac Surg. 12 (2001) 362-368. https://doi.org/10.1097/00001665-200107000-00009

[118]B.L. Schmidt, C. Lee, D.M. Young, J. O'Brien, Intraorbital squamous epithelial cyst: an unusual complication of silastic implantation. J Craniofac Surg. 9 (1998) 452-455. https://doi.org/10.1097/00001665-199809000-00012

[119]A. Laxenaire, J. Lévy, P. Blanchard, J.C. Lerondeau, F. Tesnier, P. Scheffer, Complications of silastic implants used in orbital repair (Complications des implants de silastic utilisés en réparation orbitaire). Rev Stomatol Chir Maxillofac. 98 (1997) 96-99.

[120]W.R. Dougherty, T. Wellisz, The natural history of alloplastic implants in orbital floor reconstruction: An animal model. J Craniofac Surg. 5 (1994) 26-33. https://doi.org/10.1097/00001665-199402000-00007

[121]M.J. Yaremchuk, Facial reconstruction using porous polyethylene implants. Plast Reconstr Surg. 111 (2003) 1818-1827. https://doi.org/10.1097/01.PRS.0000056866.80665.7A

[122]P.M. Villarreal, F. Monje, A.J. Morillo, L.M. Junquera, C. Gonzalez, J.J. Barbon, Porous polyethylene implants in orbital floor reconstruction. Plast Reconstr Surg. 109 (2002) 877-887. https://doi.org/10.1097/00006534-200203000-00007

[123]I. Kinnunen, K. Aitasalo, M. Pollonen, M. Varpula, Reconstruction of orbital floor fractures using bioactive glass. J Craniomaxillofac Surg. 28 (2000) 229-234. https://doi.org/10.1054/jcms.2000.0140

[124]K. Aitasalo, I. Kinnunen, J. Palmgren, M. Varpula, Repair of orbital floor fractures with bioactive glass implants. J Oral Maxillofac Surg. 59 (2001) 1390-1396. https://doi.org/10.1053/joms.2001.27524

[125]J.L. Parkin, M.H. Stevens, J.C. Stringham, Absorbable gelatin film verses silicone rubber sheeting in orbital fracture treatment. Laryngoscope. 97 (1987) 1-3. https://doi.org/10.1288/00005537-198701000-00001

[126]R.W. Mermer, R.E. Orban, Jr Repair of orbital floor fractures with absorbable gelatin film. J Craniomaxillofac Trauma. 1 (1995) 30-34.

[127]W.P. Piotrowski, U. Mayer-Zuchi, The use of polyglactin 910-polydioxanon in the treatment of defects of the orbital roof. J Oral Maxillofac Surg. 57 (1999) 1301-1306 https://doi.org/10.1016/S0278-2391(99)90865-0

[128]P.V. Hatton, J. Walsh, I.M. Brook, The response of cultured bone cells to resorbable polyglycolic acid and silicone membranes for use in orbital floor fracture repair. Clin Mater. 17 (1994) 71-80. https://doi.org/10.1016/0267-6605(94)90014-0

[129]D.J. Mackenzie, B. Arora, J. Hansen, Orbital floor repair with titanium mesh screen. J Carniomaxillofac Trauma. 5 (1999) 9-18.

[130]H.S. Park, Y.K. Kim, C.H. Yoon, Various applications of titanium mesh screen implant to orbital wall fractures. J Craniofac Surg. 12 (2001) 555-560. https://doi.org/10.1097/00001665-200111000-00010

[131]A.J. Gear, A. Lokeh, J.H. Aldridge, M.R. Migliori, C.I. Benjamin, W. Schubert, Safety of titanium mesh for orbital reconstruction. Ann Plast Surg. 48 (2002) 1-9. https://doi.org/10.1097/00000637-200201000-00001

Applications of Polymers in Surgery
Materials Research Foundations **123** (2022) 155-174

Materials Research Forum LLC
https://doi.org/10.21741/9781644901892-6

Chapter 6

Advancement of Guided Tissue Regeneration (GTR) Membranes for Dental Applications

Nayab Amin,[1,2] Usama Siddiqui[1,2] Nawshad Muhammad[1*], Saad Liaqat[1],
Muhammad Adnan Khan[1,] Abdur Rahim[3]

[1]Department of Dental Materials, Institute of Basic Medical Sciences, Khyber Medical University, Peshawar, Pakistan

[2]Department of Dental Materials, Rehman College of Dentistry, Peshawar, Pakistan

[3]Interdisciplinary Research Center for Biomedical Materials, COMSATS University Islamabad, Lahore Campus

* nawshad.ibms@kmu.edu.pk

Abstract

Periodontal disease leads to inflammation and consequently destruction of periodontium. The main aim of periodontal therapy is to minimize the inflammation by regeneration of defects. Over the past several decades, various materials as well as clinical techniques have been studied for repair and regeneration of periodontal defects. The tissue regeneration should permit proliferation of different cells in separate compartments, as four types of tissues constitute periodontium. Guided tissue regeneration (GTR) was introduced to restrict the growth of epithelium in alveolar bone space. Several barrier membranes were developed to aid in this technique which included both natural and synthetic materials. To achieve essential function of space maintenance, non-resorbable membranes were used previously with bone graft materials. Although these membranes were widely used, however they had many disadvantages which led to development of second generation (resorbable) and third generation (functional) membranes. This chapter focuses on advanced developments of biodegradable polymers for use in GTR treatment. Moreover, non-resorbable and resorbable polymeric membranes, classifications, advanced experimental research, clinical applications, and their challenges are also discussed. The composite periodontal membranes are also under research. Despite the biocompatibility and biodegradability of unreinforced barrier membranes, these are deficient in anti-bacterial properties. Therefore, barrier membranes blended with functional materials have

Applications of Polymers in Surgery   Materials Research Forum LLC
Materials Research Foundations **123** (2022) 155-174   https://doi.org/10.21741/9781644901892-6

been fabricated which exhibited more benefits. The comparison of past and recent trends of barrier membranes can guide correctly for periodontal regeneration therapy.

Keywords

Barrier Membranes, Guided Tissue Regeneration, Periodontal Biomaterials, Periodontal Regeneration

Contents

1.   Introduction

The theory of guided tissue regeneration (GTR) was first given by Melcher in 1976. According to his assumption, secretion and synthesis of cementum can be achieved from cells of periodontal ligament to attain attachment of newly formed collagen fibers to the tooth. This can be done by introducing a physical barrier under physiological circumstances. This hypothesis was later experimentally tested as well as verified histologically by Karring and his colleagues [1]. The amount of attachment can be

Materials Research Forum LLC

https://doi.org/10.21741/9781644901892-6

determined by the cells that accumulate at the root surface e.g. if migration of fibroblasts and epithelial cells occur, then bone formation can be inhibited. These effects are due to variation in the growth potential of periodontal cells [2].

The need to prevent the migration of gingival cells of both connective and epithelial tissues towards the root surface led to the advancement and use of GTR membranes. These membranes act as a barrier between gingival epithelium and alveolar bone to prevent the progression of epithelial cells into the wound. Using a barrier membrane (BM) during a GTR treatment is crucial as compartmentalizing is needed for formation of new bone and connective tissue. Besides this, it also prevents ridge collapse due to emptiness of socket [3]. The insufficiency of BM to provide occlusive as well as regenerative functions might result in issues such as excessive epithelial cell migration into the socket and inadequate bone formation. The goal of GTR treatment is to reconstruct the periodontal ligament (PDL) and restore the morphology as well as the function of the periodontium. This treatment aims to isolate the site of healing from connective tissues, stabilize the blood clot, repair the wound and offer enough space for ridge or bone healing [4].

The BM used in the GTR treatment should be in contact with soft and hard tissues and should encourage regeneration of bone. Several methods are used to fabricate BM which are electro-spinning, solvent casting, freeze-drying, phase inversion, and three-dimensional printing. The materials of the GTR barrier matrix have changed over time from non-resorbable polytetrafluoroethylene (PTFE), cellulose filters and titanium mesh to resorbable natural or synthetic polymer to eliminate the need for secondary GTR removal [5]. Commonly resorbable natural membranes are fabricated from human collage, bovine or porcine whereas resorbable synthetic membranes are made of aliphatic polyesters like polyglycolic acid (PGA), polylactic acid (PLA), and poly ε-caprolactone (PCL) or their copolymers. These membranes have been marketed by different companies under various brand names. The details of non-resorbable membranes are given in Table 1 while details of resorbable membranes are given in Table 2 [6].

Table 1: Non-resorbable Barrier Membranes Commercially Available in the Market [6]

Trade Name (Company)	Material
Cytoflex® Tefguard Unicare Biomaterial	PTFE
TefGEN-FD	PTFE
Gore-Tex®	e-PTFE
BoneShields® (FRIOS)	Titanium
Tocksystem (MeshTM)	Titanium
Ti-Micromesh (ACE)	Titanium
Cytoplast™ Ti-Reinforced (Osteogenesis Biomedical)	Ti-PTFE

Table 2: Resorbable Barrier Membranes commercially available in the market [6]

Trade Name (Company)	Material
Natural	
BioMend® (Zimmer Dental)	Bovine Collagen (Type I)
GENOSS Collagen Membrane (GENOSS)	Bovine Collagen (Type I)
Mem-Lok® Pliable (BioHorizons)	Porcine Collagen (Type I)
Kontour™ (Implant Direct)	Porcine Collagen (Type I)
Healiguide™ (Encoll)	Type I Collagen
Creos™ xenoprotect (Nobel Biocare®)	Porcine collagen (Type I & Type III)
BioGide® (Geistlich Biomaterials Switzerland)	Porcine collagen (Type I & Type III)
Mem-Lok® Pericardium (BioHorizons)	Human pericardium tissue allograft
Synthetic	
Guidor® (Sunstar)	PLA
Resolute® (Gore®)	PGA
Atrisorb (Tolmar)	PDLLA
Cytoflex Resorb® (Unicare Biomedical)	PLGA
Vivosorb® (Polyganics)	PDLLCL

2. Classification of Barrier Membranes

Barrier membranes can be classified based on:
  a) Resorb-ability (Non-resorbable and Resorbable) depicted in **Figure 1**.

  b) Generations (First, second and third) depicted in **Figure 2**.

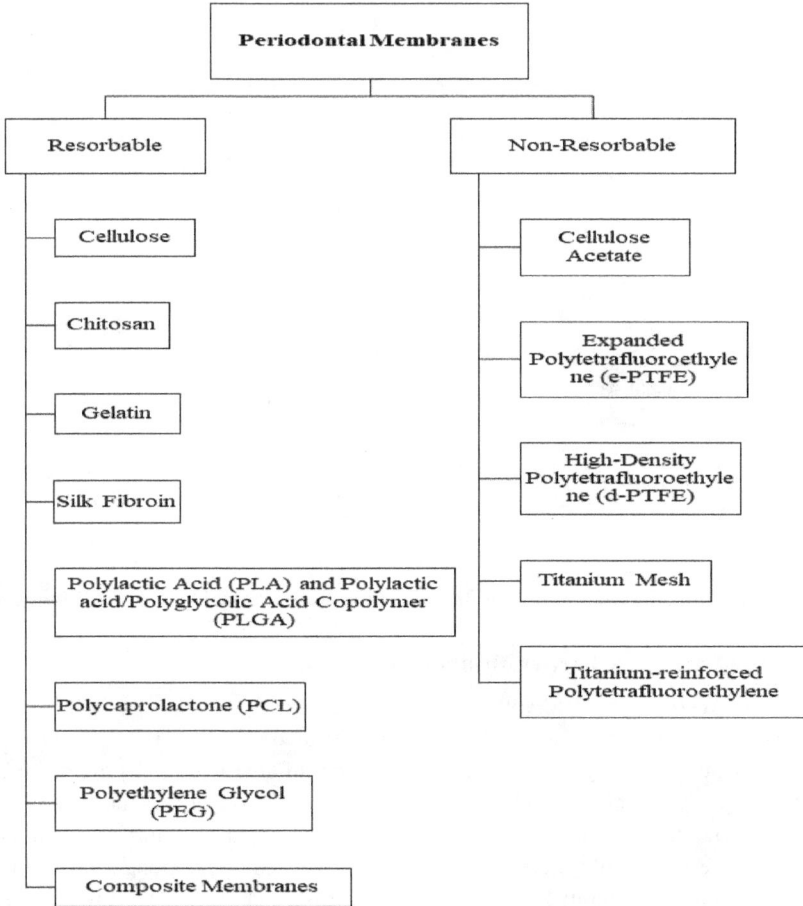

Figure 1: Classification of Barrier Membranes on basis of Resorb-ability

Applications of Polymers in Surgery   Materials Research Forum LLC
Materials Research Foundations **123** (2022) 155-174   https://doi.org/10.21741/9781644901892-6

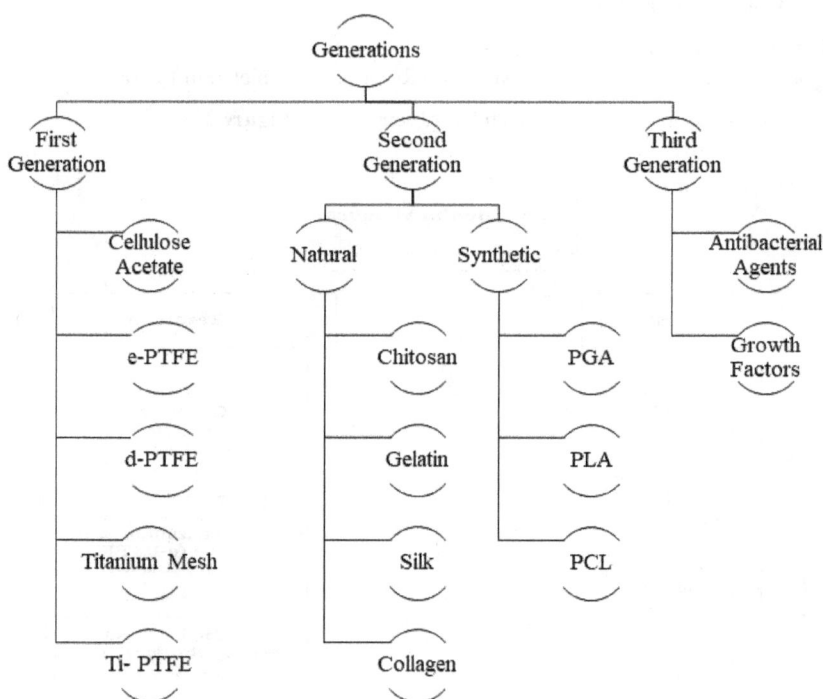

Figure 2: Classification of Barrier Membranes according to Generations

3.   Non Resorbable Barrier Membranes

Cellulose Acetate: The first synthetic material which was used as a periodontal membrane in guided tissue generation was cellulose acetate (Millipore®) [7]. It is an ester of cellulose and is manufactured by full or partial acetylation of free hydroxyl groups present in anhydro glucose component. On the basis of 29% - 48% acetyl content; they are called monoacetate, diacetate, and triacetate [8]. Due to its resistance against chemical agents, adequate thermal stability, mechanical strength, and flexibility, along with hydrophilic nature and biocompatibility [9], it was used as a periodontal membrane. However, because of poor clinical handling soon it was replaced by other materials [10].

Expanded Polytetrafluoroethylene (e-PTFE): The membranes from e-PTFE were created in 1969 and became standard for guided tissue regeneration in the early 1990s. e-

Applications of Polymers in Surgery   Materials Research Forum LLC
Materials Research Foundations **123** (2022) 155-174   https://doi.org/10.21741/9781644901892-6

PTFE is a fluorinated polymer having excellent physical and chemical properties such as resistance to deterioration under mechanical forces, non-stick properties, stability in high temperatures, and resistance to chemicals. It is synthesized by subjecting PTFE to high tensile strength and then sintering it with having pores between 5 to 20 microns in the material structure [11]. The e-PTFE membranes have 90% porous structure on one side and 30% porous structure on the other. This prevents the growth of epithelium on a more porous side during wound healing while prevention of fibrous ingrowth to provide space for new bow growth on a less porous side. The advantages of e-PTFE membranes include perseverance and regeneration of bone surrounding implants which are introduced in fresh extraction sockets while disadvantages are the need for wide releasing incisions and exposure to the oral cavity [12].

High Density Polytetrafluoroethylene (d-PTFE): Membranes from high density polytetrafluoroethylene (d-PTFE) were introduced in 1993 to overcome the complications with expanded polytetrafluoroethylene membranes. They have a pore size of 0.3 microns. This small pore size is advantageous that microorganisms cannot pass through the membrane even if it is exposed to the oral cavity [13]. Moreover, transfusion of small molecules and diffusion of oxygen across the membrane is still achievable. Therefore, even after exposure d-PTFE membranes have excellent bone regeneration results. Besides this, as no tissue ingrowth occurs into the surface of the membrane its removal is easy and simple. Additionally, d-PTFE membranes are also beneficial in cases of large bony defects, alveolar ridge preservation, and immediate implants after extraction membranes because they can be left exposed as primary closure is not achievable without tension [14]. It also improves healing as no extensive releasing incisions are needed for primary closure and help in the preservation of soft tissue and mucogingival junction. Although these membranes have many advantages, they have drawbacks in that they tend to collapse towards defect and decrease the production of collagen while increasing accumulation of glycosaminoglycan which may consequently result in delayed healing [15].

Titanium (Ti) Mesh Membrane: Titanium mesh membranes were developed in 1969. They have excellent mechanical properties such as rigidity, elasticity, wear resistance, malleability, and high strength with lightweight. As they have superior mechanical properties so they can be used in larger defects in which bone and tissue regeneration are required. Because of its low density and higher elasticity, it can be easily bent and contoured into the desired shape of the defect and inhibit mucosal compression respectively [16]. Besides this, Ti is biocompatible, it can be passivated easily to make it corrosion resistant and can be readily sterilized before implantation as it can resist high temperatures. In addition to these properties, the presence of macropores in Ti meshes is beneficial as blood supply to the tissues beneath is not directly affected. Thus, improving tissue integration and wound stability. Due to these unique properties, the Ti mesh membranes are a perfect substitute for e-PTFE as a non-resorbable membrane [17]. Despite having many advantages, these membranes have certain shortcomings such as due to their increased stiffness there are more chances of exposure of the membrane to the oral cavity due to gingival tissue perforation because of sharp ends of the membrane. Moreover, for the removal of these membranes, complex secondary surgery is required [18].

Titanium reinforced Polytetrafluoroethylene Membranes (Ti-PTFE): Titanium reinforced polytetrafluoroethylene membranes are made either by reinforcing titanium with expanded or dense polytetrafluoroethylene. The incorporation of Ti into PTFE offers a biocompatible, non-resorbable, and high strength material which is corrosion resistant material. Thus, enhancing mechanical stability, providing space and maintaining large defects, and avoiding barrier membranes from collapsing. Although Ti-PTFE membranes have benefits, however studies showed that unreinforced PTFE membranes are more advantageous as they were able to stabilize blood clots in protected large space [19, 20].

3.1   Major Disadvantages of Non-resorbable Membranes

Following are the major disadvantages of non-resorbable membranes

- Due to secondary surgery required for removal of the membrane, the cost and discomfort for the patients are increased.

- Premature contact of membranes into the oral cavity may lead to bacterial colonization and subsequently infection.

- Increase chances of development of wound dehiscence.

- Mini screws are usually needed as extra additional stabilization of non-resorbable membranes is required because of their rigidity [21, 22].

4.   Resorbable Barrier Membranes

Barrier membranes having biodegradable qualities were developed to overcome the drawbacks of non-resorbable barrier membranes. The significant advantage of a resorbable barrier membrane is that an extra surgical procedure is not required for its removal, consequently reducing cost, time, and complications. These membranes are either manufactured from natural collagen or synthetically from aliphatic esters [23].

4.1   Natural Membranes

Collagen membranes: Collagen makes up a large portion of the extracellular matrix (ECM) in the body. It offers several beneficial biological properties such as it attracts and

Applications of Polymers in Surgery   Materials Research Forum LLC
Materials Research Foundations **123** (2022) 155-174   https://doi.org/10.21741/9781644901892-6

activates gingival fibroblast cells as well as periodontal ligaments (PDL), provides hemostasis, and enhances tissue thickening. Additionally, it is biocompatible and biodegradable. Due to these characteristics, it is an excellent choice for bioresorbable barrier membranes of GTR along with a wide range of applications [24]. It is derived from the epidermis, dermis, tendon, or pericardium of humans, porcine or bovine. Among collagen types, Type I or Type III, is a major constituent of most commercially available natural resorbable membranes. Type I collagen has extensive application because it is the most abundant of all the collagens, accounting for 90 % in ECM of calcified organic bone, 80 % in connective tissue, and 25 % of the total proteins present in the body. These natural collagen membranes have a remarkable affinity towards cells, as well as bone regeneration capacity as compared to non-resorbable membranes [25].

Despite these excellent characteristics, they have several disadvantages such as concerns of disease transfer to human beings for collagen derived from animal origin, loss of ability to maintain space in a humid environment, excessively quick biodegradation, and inadequate mechanical strength. The inadequate mechanical characteristics and fast degradation of natural collagen membranes are related to high vulnerability to infection, shorter functional time, and susceptible new tissue regeneration [26]. Depending on the resorbable membrane material, the degradation timing might be rapid as compared to process of wound healing, reducing the integrity of the BM, and allowing contact with undesired epithelial cells. Collagen membrane resorption is affected by the origin of the material and the pace at which collagen is broken down into amino acid molecules and oligopeptides. The absorption of these membranes occurs via enzymatic breakdown by collagenases or proteases as well as enzymes and proteases derived from polymorphonuclear lymphocytes/macrophages and bacteria respectively [27].

Cross-linking has been in use for breaking down of the collagen membrane by chemical, physical, biological, and enzymatic mechanisms. Glutaraldehyde, poly epoxy, 1-ethyl-3-(3-dimethyl aminopropyl) carbodiimide, and diphenylphosphorylationazide are the most widely chemically utilized cross-linkers [28]. The researchers concluded from several animal and clinical investigations that the degradation rate of collagen membrane is inversely proportional the amount of cross-linking. Physical methods such as heat treatment, dehydrothermal treatment, gamma, UV, and microwave irradiations are used as alternate to efficiently produce cross-linking. Additionally, biological techniques (e.g., transglutaminase) are also employed [29].

Gelatin Membranes: Gelatin is a soluble protein that is obtained from partially denatured collagen and has gained a lot of interest due to its wide availability, ease of handling, and low cost. Its favorable characteristics, including cell adhesion and growth stimulation, immunogenicity, adhesiveness, flexibility, superior biocompatibility, and cost-effectiveness, make it a promising biomaterial for GTR [30]. However, gelatin has low mechanical strength and degrades rapidly. Cross-linking and heat treatment of gelatin by glutaraldehyde or cross-linking with N-hydroxyl succinimide and 1-ethyl-3-(3-dimethyl aminopropyl) carbodiimide are effective ways to enhance its stability and mechanical characteristics. Cross-linked fibrous gelatin membrane is rarely utilized alone for GTR because it has exceptionally low Young's modulus in humid state [31, 32].

Chitosan Membranes: Chitosan is a cationic alkaline linear polysaccharide that is chemically 1,4-linked 2-amino-2-deoxy-D-glucan. It is produced through chitin deacetylation and is an excellent material for GTR membranes owing to its suitable degradation rate, hemostatic ability, potential for wound healing, non-antigenicity, flexibility in wet conditions, anti-microbial activity, high biocompatibility, and cost-efficiency. Chemical cross-linking is an efficient approach for increasing its mechanical strength and reducing degradation rate [33, 34].

Silk Fibroin Membranes: Silk fibroin is a natural protein that is usually obtained from spiders or silkworms. It has several appealing characteristics, such as oxygen and water vapor permeability, less inflammatory reaction, good biocompatibility, and biodegradability. Thus, it has been considered as a candidate material for periodontal and bone regenerative applications [35]. Silk fibroin has exceptional strength and toughness; therefore, it offers enough stability that helps in maintaining space for bone ingrowth and prevention of membrane collapse. The research showed the bone development of the silk fibroin membrane was greater than that of PTFE and cross-linked collagen, as determined by histological examination and micro computed tomography. Likewise, it has greater tensile strength as compared to above mentioned two membranes in moist conditions. All these findings imply that the silk fibroin membrane might be beneficial as a barrier membrane for GTR [36].

4.2   Synthetic Membranes

Polylactic Acid (PLA), Polyglycolic Acid (PGA), and Copolymers: Polylactic Acid (PLA) and polyglycolic acid or polyglycolide (PGA) are widely used polymers in GTR techniques. Similarly, their copolymers and blends have been in use for tissue regeneration. Guidor® Matrix Barrier is the first widely used blend in barrier technology that is prepared from poly-L-lactide (PLLA) and poly-D, L-lactide (PDLLA), forms of PLA, doped with acetyl tri-n-butyl citrate [37]. It functions as a barrier for at least 6 months and resorbs slowly in 13 months. Both PLA and PGA are synthetically prepared and biodegradable materials with good characteristics including high mechanical strength as well as elastic modulus and better performance. Most manufacturing techniques produced rigid PLA membranes, which hampered their medical uses; however, this disadvantage may be overcome by the addition of softeners as N-methyl-2-pyrrolidone (NMP) [38]. Glycolide, ε-caprolactone, and lactide copolymers have been also synthesized for the regulation of hydrophilicity and degradation rate of PLA. Proteolytic enzymes, from the polymorphonuclear (PMN) cells, degrade PLA and PGA into lactic acid and glycolic acid respectively. Consequently, they are either expelled via kidney or utilized as pyruvate in Kreb's cycle. These cells also play an important role in the inflammation process and frequently produce damaging oxidative species when breaking down synthetic membranes [39].

Polycaprolactone (PCL): Polycaprolactone (PCL) is a synthetically prepared polymer that has several attributes that meet the GTR membrane criteria. These include cost-effectiveness, high mechanical strength, and non-toxicity. Moreover, it also elicits less immune response and does not generate an excessively acidic environment during the breakdown process [40]. Besides these, it has disadvantages such as slow degradation and hydrophobicity. The degradation of PCL occurs so slowly that the total bioresorption time in *vivo* is around 2–3 years that is lengthy to be used in the treatment of GTR. Furthermore, the plasma treatment of PCL is done to overcome its hydrophobicity as this reduces adhesion of cells and proliferation [41].

Polyethylene Glycol (PEG): Polyethylene glycol (PEG) has been proposed as a possibility for GTR membranes as a cell-occlusive, biodegradable, and biocompatible polymer. *In-vivo* studies demonstrated that PEG hydrogel has good tissue integration and biocompatibility, but degradation was influenced by PEG composition. Furthermore, recent research revealed that these membranes have the potential to be used for phase amplification in intimidating lateral ridge deficiencies while preserving the ridge contours [42].

4.3   Composite Membranes

The natural and synthetic resorbable periodontal membranes meet many requirements which are needed for guided tissue regeneration. These include sufficient physical and mechanical properties, adequate strength to prohibit collapse of the membrane, biocompatibility, appropriate degradation mechanisms, acceptable barrier function, and no need for secondary surgery. Although resorbable membranes have these properties, however, a single membrane still does not meet the criteria needed for an ideal membrane. Moreover, a resorbable periodontal membrane degrades rapidly as compared to deposition of newly underlying tissue thus resulting in immature tissue as before development of true tissue has not occurred yet. Therefore, composite membranes were developed which are formed by the combination of two or more membranes/materials. These membranes are made to address the limitations of individual membranes and to accomplish positive properties of both in a synergetic way. They are continuously modified to achieve the best possible periodontal regeneration membrane [43, 44].

On basis of bioactivity, composite membranes can be classified as

- Non-functional Membranes
- Functional Membranes

5.   Non-Functional Membranes

Non-functional membranes are formed by the combination of materials that do not elicit any favorable biological response. These membranes are made either by blends of natural materials, a combination of synthetic materials, or a combination of natural and synthetic materials.

Blends of Natural Materials: As stated previously, chitosan is a natural polymer and is used in GTR due to its advantageous properties however, it has poor mechanical properties. Another, natural material that is vastly used as a membrane is GTR is gelatin. Gelatin has a limitation of high degradation rate, so a blend of chitosan and gelatin was formed to overcome the disadvantages of both. This blend is easy to make as gelatins have free carboxyl groups in their backbone structure which form a hydrogen bond with chitosan. This blend showed better cell adhesion and proliferation properties and mechanical properties as compared to chitosan or gelatin used alone [45]. Besides, this proanthocyanidin is also added in this blend which forms a more stable membrane with high mechanical properties as compared to gelatin/chitosan membranes [46].

Blends of Synthetic Materials: Polylactic acid/polyglycolic acid copolymer (PLGA) is renowned material for having decent impact on the repair of different tissues due to its better cytological compatibility. For GTR, a variety of PDLLA/PLGA electrospun membrane systems have been developed which have suitable degradation rates and good cell occlusives. Studies showed that the composite membranes of PDLLA/PLGA can serve as a physical membrane to avoid infiltration of cells. So, these composite membranes have the potential to be an effective BM for GTR [47]. Polycaprolactone is also blended with PLGA and PLA for the preparation of synthetic GTR membranes. Besides these several other synthetic composite polymers may also have a promising future in GTR [48].

Blends of Natural and Synthetic Materials: Natural materials are always more biocompatible and have more bioactive properties. Gelatin, for example, has a high number of integrin-binding sites, which are required for the adhesion and differentiation of cells. It is blended with polycaprolactone to obtain a new material with better biocompatibility and enhanced properties. Additionally, the time for biodegradation is acceptable for tissue regeneration. The PCL-Gelatin blend has been successfully in use for GTR. However, the fabrication of this composite with outstanding results is difficult due to phase separation. So, to overcome this problem acetic acid is usually added to the blend as it enhances homogeneity by mediating miscibility [49].

Likewise, PLLA-Chitosan electrospun hybrid membranes exhibited more potentiality for clinical uses as compared to unreinforced PLLA. Moreover, this blend had the property of non-fibroblast penetration as well as degradation rate was faster. Also, the multiphasic membrane of PLLA-Chitosan was prepared in a way that the outer layer was of chitosan mesh whereas the inner layer was of nano-porous PLLA to achieve enough mechanical strength. This composite had better integrity and a slow degradation rate [50].

6.   Functional Membranes

Different materials are added to impart bioactivity in the barrier membranes. These include antibacterial agents such as metronidazole [51], tetracycline [52], etc., and growth factors e.g., TGF-1 [53], platelet granules [54], and many others [55].

Anti-bacterial Agents: Controlling bacterial inflammation in the periodontium is critical for effective regeneration during the nonsurgical stage as well as the surgical stage. In addition, bacterial infections are now considered to be the most prevalent cause of GTR membrane failure in clinical applications. The research showed that 25% by weight of metronidazole benzoate may be introduced in the layer nearby epithelial tissue to inhibit the growth of bacteria and development of biofilm. Likewise, the membranes containing metronidazole significantly improved regeneration of periodontal tissues and bone in GTR treatment [3]. In another study, the PCL/gelatin electrospun membrane containing metronidazole showed improved antibacterial capability and biocompatibility. The barrier membranes made up of PLA and PCL were also incorporated with metronidazole for the anti-bacterial solution of GTR [55].

Another drug, tetracycline hydrochloride, which is likewise effective against periodontal infections, is also added into PLA and polyethylene-co-vinyl acetate membranes. The incorporation of tetracycline or amoxicillin into GTR membranes has been shown to improve periodontal ligament cell attachment in the existence of oral microbes such as *Aggregatibacter Actinomycetemcomitans* and *Streptococcus mutans*. Tetracycline has a longer release duration in collagen membranes as compared to other antibiotics, thus it might be utilized in situations where the membrane has to be used for an extended length of time [56].

Other than antibiotics, antibacterial agents which are non-antibiotic offer greater antibacterial activity and have been utilized in GTR. Chlorhexidine and chitosan nanoparticles were also combined with collagen membranes to provide antibacterial activity for periapical GTR. Chitosan nanoparticles can act synergistically with chlorhexidine to promote periapical GTR in collagen membranes. So, the anti-microbial property of chitosan may be employed to enhance regeneration techniques in periapical surgery [57].

Growth factors: Growth factors play an important role in the formation and repair of tissues, the healing process, chemotaxis, cell proliferation, and angiogenesis. These are important signaling molecules that direct cell activity by attaching to particular transmembrane receptors present on target cells. So, tissue regeneration can be accomplished by allowing control delivery of growth factors. Some bioactive molecules, including TGF-1, PDGF, EMD, and BMP-2 have been demonstrated to promote periodontal regeneration [58].

Platelet granules have many growth factors that aid in the repair mechanisms of soft and hard tissues. The compressed form of platelet-rich fibrin (PRF) has been used as BM in GTR treatment as it is relatively safe, cost-effective, and has autologous nature. It has biopolymer fibrin that serves as the foundation of growth factors facilitating tissue regeneration. However, it rapidly degrades in 2 weeks. The cross-linking of PRF may make it suitable for GTR treatment by balancing its adequate bioactivity and degradability time. The heat treatment while preparing PRF also slowed down the degradation rate without affecting biocompatibility. Therefore, the PRF membrane has a promising future as a BM in GTR treatment [59].

Conclusion

The results of advanced barrier membranes are encouraging; however due to inadequate understanding of tissue regeneration mechanisms improvement in novel science models is required. The barrier membranes containing functional biomaterials mimic native ECM closely, therefore, these could succeed in periodontal GTR therapy. As each BM has benefits and drawbacks, so the selection of BM should be based on a comprehensive understanding of advantages and limitations of the materials. This is essential to harmonize with functional needs and clinical applications.

References

[1] Satpathy A, Mohanty R, Rautray TR. Thin membrane with biomimetic hexagonal patterned surface for guided bone regeneration. Int. J. Nano Biomater. 7 (2018) 275-281. https://doi.org/10.1504/IJNBM.2018.10016280

[2] Novaes Jr AB, Palioto DB, Andrade PFd, Marchesan JT. Regeneration of class II furcation defects: determinants of increased success. Braz Dent J. 16 (2005) 87-97. https://doi.org/10.1590/S0103-64402005000200001

[3] Bottino MC, Thomas V, Schmidt G, Vohra YK, Chu T-MG, Kowolik MJ, et al. Recent advances in the development of GTR/GBR membranes for periodontal regeneration-a materials perspective. Dent. Mater. 28 (2012) 703-721. https://doi.org/10.1016/j.dental.2012.04.022

[4] Bottino MC, Thomas V. Membranes for periodontal regeneration-a materials perspective. J. Funct. Biomater. 17 (2015) 90-100. https://doi.org/10.1159/000381699

[5] Iviglia G, Kargozar S, Baino F. Biomaterials, current strategies, and novel nano-technological approaches for periodontal regeneration. J. Funct. Biomater. 10 (2019) 2-36. https://doi.org/10.3390/jfb10010003

[6] Rodriguez I, Selders G, Fetz A, Gehrmann C, Stein S, Evensky J, et al. Barrier membranes for dental applications: A review and sweet advancement in membrane developments. Mouth Teeth. 2 (2018) 1-9. https://doi.org/10.15761/MTJ.1000108

[7] Melcher A. On the repair potential of periodontal tissues. J. Periodontol. 47 (1976) 256-260. https://doi.org/10.1902/jop.1976.47.5.256

[8] Alfassi G, Rein DM, Shpigelman A, Cohen Y. Partially Acetylated Cellulose Dissolved in Aqueous Solution: Physical Properties and Enzymatic Hydrolysis. Polymers. 11 (2019) 1734. https://doi.org/10.3390/polym11111734

[9] Oprea M, Voicu SI. Recent advances in applications of cellulose derivatives-based composite membranes with hydroxyapatite. Materials. 13 (2020) 2481. https://doi.org/10.3390/ma13112481

[10] Ausenda F, Rasperini G, Acunzo R, Gorbunkova A, Pagni G. New perspectives in the use of biomaterials for periodontal regeneration. Materials. 12 (2019) 2197. https://doi.org/10.3390/ma12132197

[11] Liu J, Kerns DG. Suppl 1: Mechanisms of guided bone regeneration: A review. Open Dent. J. 8 (2014) 56. https://doi.org/10.2174/1874210601408010056

[12] Zhang Y, Zhang X, Shi B, Miron R. Membranes for guided tissue and bone regeneration. Ann. Maxillofac. Surg. 1 (2013)10. https://doi.org/10.13172/2052-7837-1-1-451

[13] Madhuri SV. membranes for Periodontal Regeneration. Int. J. Pharm. Sci. Invent. 5 (2016) 19-24.

[14] Marouf HA, El-Guindi HM. Efficacy of high-density versus semipermeable PTFE membranes in an elderly experimental model. Oral Surg. Oral Med. Oral Pathol. Oral Radiol. and Endod. 89 (2000) 164-170. https://doi.org/10.1067/moe.2000.98922

[15] Hoffmann O, Bartee BK, Beaumont C, Kasaj A, Deli G, Zafiropoulos GG. Alveolar bone preservation in extraction sockets using non-resorbable dPTFE membranes: a retrospective non-randomized study. J. Periodontol. 79 (2008) 1355-1369. https://doi.org/10.1902/jop.2008.070502

[16] Rakhmatia YD, Ayukawa Y, Furuhashi A, Koyano K. Current barrier membranes: titanium mesh and other membranes for guided bone regeneration in dental applications. J. Prosthodont. Res. 57 (2013) 3-14. https://doi.org/10.1016/j.jpor.2012.12.001

[17] Toygar HU, Guzeldemir E, Cilasun U, Akkor D, Arpak N. Long-term clinical evaluation and SEM analysis of the e-PTFE and titanium membranes in guided tissue regeneration. J. Biomed. Mater. Res. 91 (2009) 772-779. https://doi.org/10.1002/jbm.b.31454

[18] Li J, Hua L, Wang W, Gu C, Du J, Hu C. Flexible, high-strength titanium nanowire for scaffold biomimetic periodontal membrane. Biosurf and Biotribol. 7 (2021) 23-29. https://doi.org/10.1049/bsb2.12010

[19] Cucchi A, Ghensi P. Vertical guided bone regeneration using titanium-reinforced d-PTFE membrane and prehydrated corticocancellous bone graft. Open Dent. J. 8 (2014) 194. https://doi.org/10.2174/1874210601408010194

[20] Windisch P, Orban K, Salvi GE, Sculean A, Molnar B. Vertical-guided bone regeneration with a titanium-reinforced d-PTFE membrane utilizing a novel split-thickness flap design: a prospective case series. Clin. Oral Investig. 25 (2021) 2969-2980. https://doi.org/10.1007/s00784-020-03617-6

[21] Soldatos NK, Stylianou P, Koidou VP, Angelov N, Yukna R, Romanos GE. Limitations and options using resorbable versus nonresorbable membranes for successful guided bone regeneration. Quintessence Int. 48 (2017) 131-147.

[22] Naung NY, Shehata E, Van Sickels JE. Resorbable versus nonresorbable membranes: when and why? Dent. Clin. N. Am. 63 (2019) 419-431. https://doi.org/10.1016/j.cden.2019.02.008

[23] Jo Y-Y, Oh J-H. New resorbable membrane materials for guided bone regeneration. Appl. Sci. 8 (2018) 2157. https://doi.org/10.3390/app8112157

[24] Stoecklin-Wasmer C, Rutjes A, Da Costa B, Salvi G, Jüni P, Sculean A. Absorbable collagen membranes for periodontal regeneration: a systematic review. J. Dent. Res. 92 (2013) 773-781. https://doi.org/10.1177/0022034513496428

[25] Behring J, Junker R, Walboomers XF, Chessnut B, Jansen JA. Toward guided tissue and bone regeneration: morphology, attachment, proliferation, and migration of cells cultured on collagen barrier membranes. A systematic review. Odontology. 96 (2008) 1-11. https://doi.org/10.1007/s10266-008-0087-y

[26] Bunyaratavej P, Wang HL. Collagen membranes: a review. J. Periodontol. 272 (2001) 215-229. https://doi.org/10.1902/jop.2001.72.2.215

[27] Sbricoli L, Guazzo R, Annunziata M, Gobbato L, Bressan E, Nastri L. Selection of collagen membranes for bone regeneration: a literature review. Materials. 13 (2020) 786. https://doi.org/10.3390/ma13030786

[28] Wang J, Wang L, Zhou Z, Lai H, Xu P, Liao L, et al. Biodegradable polymer membranes applied in guided bone/tissue regeneration: a review. Polymers. 8 (2016) 115. https://doi.org/10.3390/polym8040115

[29] Caballé-Serrano J, Sawada K, Miron RJ, Bosshardt DD, Buser D, Gruber R. Collagen barrier membranes adsorb growth factors liberated from autogenous bone chips. Clin Oral Implants Res. 28 (2017) 236-241. https://doi.org/10.1111/clr.12789

[30] Isik G, Hasirci N, Tezcaner A, Kiziltay A. Multifunctional periodontal membrane for treatment and regeneration purposes. J. Bioact. Compat. Polym. 35 (2020) 117-138. https://doi.org/10.1177/0883911520911659

[31] Zhang S, Huang Y, Yang X, Mei F, Ma Q, Chen G, et al. Gelatin nanofibrous membrane fabricated by electrospinning of aqueous gelatin solution for guided tissue regeneration. J. Biomed. Mater. Res. A. 90 (2009) 671-679. https://doi.org/10.1002/jbm.a.32136

[32] Chou J, Komuro M, Hao J, Kuroda S, Hattori Y, Ben-Nissan B, et al. Bioresorbable zinc hydroxyapatite guided bone regeneration membrane for bone regeneration. Clin. Oral Implants Res. 27 (2016) 354-360. https://doi.org/10.1111/clr.12520

[33] Xu C, Lei C, Meng L, Wang C, Song Y. Chitosan as a barrier membrane material in periodontal tissue regeneration. J. Biomed. Mater. Res. B. 100 (2012) 1435-1443. https://doi.org/10.1002/jbm.b.32662

[34] Qasim SB, Najeeb S, Delaine-Smith RM, Rawlinson A, Rehman IU. Potential of electrospun chitosan fibers as a surface layer in functionally graded GTR membrane

for periodontal regeneration. Dent. Mater. 33 (2017) 71-83.
https://doi.org/10.1016/j.dental.2016.10.003

[35] Geão C, Costa-Pinto AR, Cunha-Reis C, Ribeiro VP, Vieira S, Oliveira JM, et al. Thermal annealed silk fibroin membranes for periodontal guided tissue regeneration. J. Mater. Sci. Mater. Med. 30 (2019) 27. https://doi.org/10.1007/s10856-019-6225-y

[36] Ha Y-Y, Park Y-W, Kweon H, Jo Y-Y, Kim S-G. Comparison of the physical properties and in vivo bioactivities of silkworm-cocoon-derived silk membrane, collagen membrane, and polytetrafluoroethylene membrane for guided bone regeneration. Macromol. Res. 22 (2014)1018-1023. https://doi.org/10.1007/s13233-014-2138-2

[37] Kim EJ, Yoon SJ, Yeo G-D, Pai C-M, Kang I-K. Preparation of biodegradable PLA/PLGA membranes with PGA mesh and their application for periodontal guided tissue regeneration. Biomed. Mater. 4 (2009) 055001. https://doi.org/10.1088/1748-6041/4/5/055001

[38] Gentile P, Chiono V, Tonda-Turo C, Ferreira AM, Ciardelli G. Polymeric membranes for guided bone regeneration. Biotechnol. J. 6 (2011) 1187-1197. https://doi.org/10.1002/biot.201100294

[39] Budak K, Sogut O, Sezer UA. A review on synthesis and biomedical applications of polyglycolic acid. J. Polym. Res. 27 (2020) 1-19. https://doi.org/10.1007/s10965-020-02187-1

[40] Osathanon T, Chanjavanakul P, Kongdecha P, Clayhan P, Huynh NC-N. Polycaprolactone-Based Biomaterials for Guided Tissue Regeneration Membrane. Periodontitis-A Useful Reference: Intech Open. (2017) 171-188. https://doi.org/10.5772/intechopen.69153

[41] Osorio R, Alfonso-Rodríguez CA, Osorio E, Medina-Castillo AL, Alaminos M, Toledano-Osorio M, et al. Novel potential scaffold for periodontal tissue engineering. Clin Oral Investig. (2017) 2695-2707. https://doi.org/10.1007/s00784-017-2072-8

[42] Yazdi MK, Vatanpour V, Taghizadeh A, Taghizadeh M, Ganjali MR, Munir MT, et al. Hydrogel membranes: A review. Mater. Sci. and Eng. C. 114 (2020) 111023. https://doi.org/10.1016/j.msec.2020.111023

[43] Zupancic S, Kocbek P, Baumgartner S, Kristl J. Contribution of nanotechnology to improved treatment of periodontal disease. Curr. Pharm. Des. 21 (2015) 3257-3271. https://doi.org/10.2174/1381612821666150531171829

[44] Sunandhakumari VJ, Vidhyadharan AK, Alim A, Kumar D, Ravindran J, Krishna A, et al. Fabrication and in vitro characterization of bioactive glass/nano hydroxyapatite reinforced electrospun poly (ε-caprolactone) composite membranes for guided tissue regeneration. Bioengineering. 5 (2018) 54. https://doi.org/10.3390/bioengineering5030054

[45] Shen R, Xu W, Xue Y, Chen L, Ye H, Zhong E, et al. The use of chitosan/PLA nano-fibers by emulsion eletrospinning for periodontal tissue engineering. Artif Cells Nanomed Biotechnol. 46 (2018) 419-430. https://doi.org/10.1080/21691401.2018.1458233

[46] Kim SW. The study of chitosan/gelatin based films crosslinked by proanthocyanidins as biomaterials: University of Southern California; 2004.

[47] Zhang E, Zhu C, Yang J, Sun H, Zhang X, Li S, et al. Electrospun PDLLA/PLGA composite membranes for potential application in guided tissue regeneration. Mater. Sci. Eng. C. 58 (2016) 278-285. https://doi.org/10.1016/j.msec.2015.08.032

[48] Agarwal S, Wendorff JH, Greiner A. Use of electrospinning technique for biomedical applications. Polymer. 49 (2008) 5603-5621. https://doi.org/10.1016/j.polymer.2008.09.014

[49] Zhang L, Dong Y, Zhang N, Shi J, Zhang X, Qi C, et al. Potentials of sandwich-like chitosan/polycaprolactone/gelatin scaffolds for guided tissue regeneration membrane. Mater. Sci. Eng. C. 109 (2020) 110618. https://doi.org/10.1016/j.msec.2019.110618

[50] Chen S, Hao Y, Cui W, Chang J, Zhou Y. Biodegradable electrospun PLLA/chitosan membrane as guided tissue regeneration membrane for treating periodontitis. J. Mater. Sci. 48 (2013) 6567-6577. https://doi.org/10.1007/s10853-013-7453-z

[51] Ardhani R, Hafiyyah OA, Ana ID, editors. Preparation of carbonated apatite membrane as metronidazole delivery system for periodontal application. Key Eng. Mater. 696 (2016) 250-258. https://doi.org/10.4028/www.scientific.net/KEM.696.250

[52] Lee SB, Lee DY, Lee YK, Kim KN, Choi SH, Kim KM. Surface modification of a guided tissue regeneration membrane using tetracycline-containing biodegradable polymers. Surf. Interface Anal. 40 (2008):192-197. https://doi.org/10.1002/sia.2761

[53] Ochiai H, Yamamoto Y, Yokoyama A, Yamashita H, Matsuzaka K, Abe S, et al. Dual nature of TGF-β1 in osteoblastic differentiation of human periodontal ligament cells. J. Hard Tissue Biol. 19 (2010) 187-194. https://doi.org/10.2485/jhtb.19.187

[54] Navarro LB, Barchiki F, Junior WN, Carneiro E, da Silva Neto UX, Westphalen VPD. Assessment of platelet-rich fibrin in the maintenance and recovery of cell

viability of the periodontal ligament. Sci Rep. 9 (2019) 1-9.
https://doi.org/10.1038/s41598-019-55930-0

[55] Nityasri AS, Pradeep K. Role of CGF (Concentrated Growth Factor) in periodontal
regeneration. J. Dent. Health Oral Disord Ther. 9 (2018) 350-352.
https://doi.org/10.15406/jdhodt.2018.09.00407

[56] Hung S-L, Lin Y-W, Chen Y-T, Ling L-J. Attachment of periodontal ligament cells
onto various antibiotics-loaded GTR membranes. Int. J. Periodontics Restorative Dent.
25 (2005) 265-275

[57] Barreras US, Méndez FT, Martínez REM, Valencia CS, Rodríguez PRM, Rodríguez
JPL. Chitosan nanoparticles enhance the antibacterial activity of chlorhexidine in
collagen membranes used for periapical guided tissue regeneration. Mater. Sci. Eng. C.
Mater. 58 (2016) 1182-1187. https://doi.org/10.1016/j.msec.2015.09.085

[58] Sam G, Pillai BRM. Evolution of barrier membranes in periodontal regeneration-
"are the third generation membranes really here? J. Clin. Diagn. Res. 8 (2014) 14-17.
https://doi.org/10.7860/JCDR/2014/9957.5272

[59] Preeja C, Arun S. Platelet-rich fibrin: Its role in periodontal regeneration. Saudi J.
Dent Res. 5 (2014) 117-122. https://doi.org/10.1016/j.ksujds.2013.09.001

Applications of Polymers in Surgery   Materials Research Forum LLC
Materials Research Foundations **123** (2022) 175-194   https://doi.org/10.21741/9781644901892-7

Chapter 7

Advancements in Polymeric Denture Lining Materials

Ghazal Khan, Khumara Roghani, Nawshad Muhammad*, Saad Liaqat, Muhammad Adnan Khan

Department of Dental Materials, Institute of Basic Medical Sciences, Khyber Medical University, Peshawar, Pakistan

*nawshad.ibms@kmu.edu.pk

Abstract

Denture liners are elastomeric materials that are applied on the fitting surface of a partial or complete removable denture. They are mainly used to improve the fit of dentures and provide a cushioning effect, to relieve pain and discomfort of the patient. Denture liners are broadly categorized into three main groups (1) Soft liners (2) Tissue Conditioners and (3) Hardliners and can be either heat cured, light cured, or self-cured. Although these materials have advantageous properties each one of these has its disadvantages as well. Therefore, the selection of material for a specific condition is of great importance and depends on the properties of the selected material. In this chapter, we will discuss the different types of denture liners, their clinical application, properties, limitations, and recent advances in these denture liners.

Keywords

Soft Denture Liner, Tissue Conditioner, Adhesion, Denture Stomatitis

Contents

1.   Introduction

Prosthodontics is as defined by Glossary of Prosthodontics as," dental specialty pertaining to the diagnosis, treatment planning, rehabilitation and maintenance of oral function, comfort, appearance and health of patients with clinical conditions associated with missing or deficient teeth and/or maxillofacial tissues using biocompatible substitutes [1]"

Over a wide variety of treatment options provided to patients by prosthodontists, one of them is Denture. A type of dental prosthesis that is used to replace the lost teeth is known as Denture. It can either be a partial denture that replaces some of the lost teeth or a complete/full denture that is a removable replacement for all of the missing teeth. A typically removable denture has two main parts; Denture Base and Artificial Teeth. Generally, denture base is made from acrylic resin, though this material can last for years, they have certain limitations. Leaching of monomer causes porosity and microbial adhesion causing denture stomatitis. Another problem is ridge resorption in patients with time leads to loose dentures which is the cause of pain for the patient. To overcome these problems Denture liners were introduced.

Denture liners are the materials that are applied on the denture base of an already fabricated denture for several reasons. It is mainly indicated for use in ill-fitting dentures to improve its fit leading to even distribution of masticatory forces and to relieve pain due to soreness.

Applications of Polymers in Surgery   Materials Research Forum LLC
Materials Research Foundations **123** (2022) 175-194   https://doi.org/10.21741/9781644901892-7

Sometimes the denture base cannot be tolerated by patients and therefore a soft barrier is required. In such cases, a type of liner is used that gives a cushioning effect and prevents irritation due to denture. In some cases, it is required for producing a functional impression.

Denture lining is a less costly and less time consuming procedure compared to fabricating a new denture thereby their use has increased considerably, but despite that these denture lining materials still possess some major disadvantages like debonding from denture base, fungal growth leading to stomatitis, and degradation in the oral environment. New preparations of denture liners are being studied to limit the disadvantages of denture liners.

Denture liners are of various types (heat cured, light cured, and self-cured) and have a different composition (silicone-based and acrylic-based) and are broadly classified into three main groups that are soft lining materials, Tissue Conditioners and Hard reliners.

2. Types of denture liners

2.1 Soft liners

A soft liner is a type of lining material that is most frequently used to reline a denture [2]. The term "soft liners" refers to a group of resilient dental materials that are applied on the fitting surfaces of the denture to form a soft padded layer between the hard denture base and the contacting occlusal surfaces of the oral mucosa [3]. Application of a liner is relatively a noninvasive and cheaper process than fabricating a new denture [4].

Soft liners help to ease discomfort, soreness, and pain from dentures [5]. They are known to provide relief to denture wearing patients, who present with severe ridge resorption, bony undercuts, and non-resilient or thin oral mucosa [3]. Soft liners act as shock absorbers for the underlying residual ridge and are effective in minimizing the impact forces compared with a hard denture base alone.

Soft liners are made resilient by co-polymerization with the monomer or by adding alcohol type plasticizers to their composition [5]. They can either be Heat polymerized when their processing is carried out in a dental laboratory [6] or Self polymerized when they are processed in a dentist clinic [7]. They are available in different compositions, which determine the properties, indications, and durability of the material [8].

2.1.1 Types of soft denture liners

O'Brien classified soft liners, based on their composition, as silicone-based or acrylic-based [9]. Based on their durability, soft liners can also be classified as short-term (30 days) and long-term soft liners (more than 30 days) [10, 11]. However, most of the literature supports the idea that temporary liners can serve for 6 months while permanent liners can serve for a year. Long term soft liners are further classified into four groups based on chemical structure:

1) Self-cured silicone 2) heat-cured silicone 3) Self-cured acrylic resin 4) heat-cured acrylic resin (Fig.1)

The acrylic-based soft liners have viscoelastic properties while silicone-based soft liners show elastic properties [12] and are comparatively long-lasting and mechanically superior to resin-based liners [13].

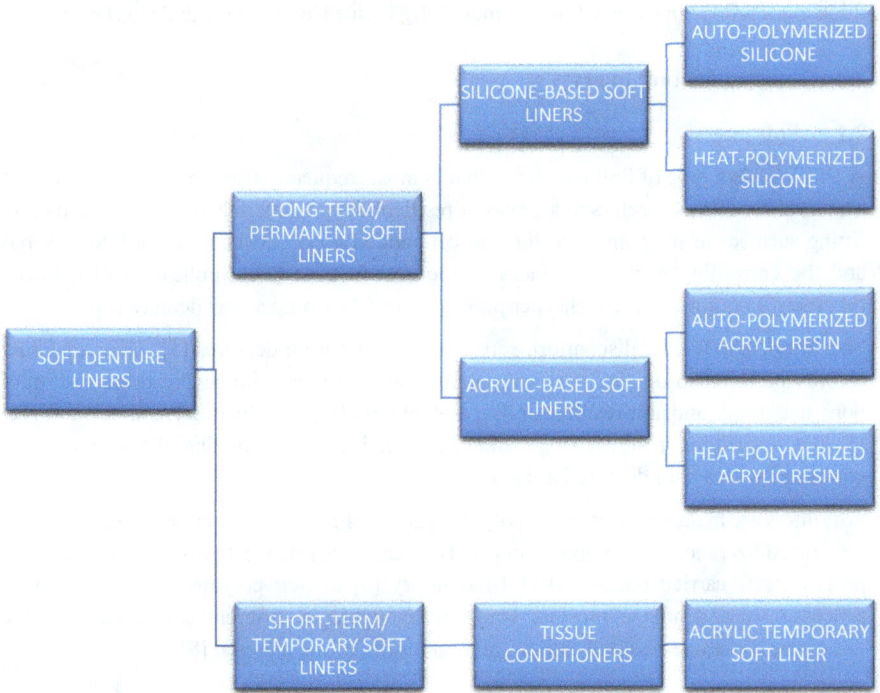

Fig. 1 classification of soft denture liners taken with permission from publisher (11)

2.1.2 Drawbacks of soft denture liners

Failures in the physical properties of the soft lining material mainly result in its poor clinical performance [14]. Clinical limitations of soft liners include the presence of surface defects and porosity [15], poor adhesion to acrylic resin [16], water uptake [17], a tendency to color change [18], and difficulty in cleaning [19].

Surface roughness

Rough surfaces facilitate microbial growth resulting in denture stomatitis and affect the durability of the lining material [20]. Surface roughness can be caused by certain denture cleaning solutions. Such solutions are known to change the morphology of lining material by penetrating the resin. Also, the concentration and immersion time can alter the polymer structure [21]. Literature also suggests that the denture type, habits like smoking, nighttime denture wearing, and pH of resting saliva cause worsening of acrylic soft liners [22].

To prevent the surface changes in soft liners, Singh K et al. suggested that coating can be done to reduce the loss of surface integrity of soft lining material [23]. Most of the literature suggests that the type of coating material used can also affect their performance in maintaining surface softness and preventing surface porosities [21, 24]. The sealed soft liners are also preferred because they stay resilient, clean and prevent the formation of biofilm and microbial growth for a long time as compared to non-sealed materials [8, 25].

Adhesion/ bond strength

For the denture to function clinically, a strong bond between the denture base and soft liner is necessary. To ensure strong bonding between the denture base and the lining material, the monomer must enter and infuse the resin pores to form an interpenetrating polymeric complex [26]. Aging is also a common factor that adversely affects the interface between the denture base and the lining material and changes its adhesive properties [27]. Soft lining materials absorb oral fluids and moisture which causes the material to swell up, resulting in stress development between the bonding surfaces, hence leading to an adhesive failure [28].

In literature, many procedures are suggested to increase the bond strength of the lining materials. For this reason, different bonding agents have been used. Lassila et al. in 2010 used ethyl acetate as a bonding agent and found better adhesion between denture base resin and lining material [29]. Primers and adhesives prevent bubble formation during denture reline procedures. Therefore, they are also used to stick silicone liners to denture base resin [30]. The use of organic solvents (MMA and ethyl acetate) result in porosities that improve monomer penetration, hence improving bond strength [31, 32]. Lasers and alumina abrasion have also been tested to overcome the clinical limitation of de-bonding [33].

Water sorption

Another common drawback of soft liners is their ability to absorb water and solubility. It is related to the changes in morphology as well as the physical properties of the lining material. These changes can lead to swelling, wear, fungal growth, and stress accumulation at the bonding surfaces [34].

Water sorption is due to the hydrophilic and porous nature of the material and is affected by the presence of cross linking agents [35]. The use of sodium perborate as a chemical denture cleaning method can enhance the water sorption of soft lining materials [36]. Barbara et al. concluded that tissue conditioners coated with certain protective preparations such ass monopoly did not absorb water in vitro [37]. One of the studies claimed that the type of filler added in soft liner composition causes water absorption [3]. In 1996, Waters also proved from his research that heat polymerization of silicone led to the production of a denser material which was lacking micro pockets of fluid/water [38].

Color stability

Color stability is another property of the lining material for maintaining the quality of the material [39]. It can be compromised by a variety of denture cleansers and by immersion time [40]. Color changes may also result from aging, staining from nicotine containing beverages like tea and coffee (Fig. 2), and microbial colonization [41]. Ideally, a denture soft liner should not easily undergo color change or staining after long use [39].

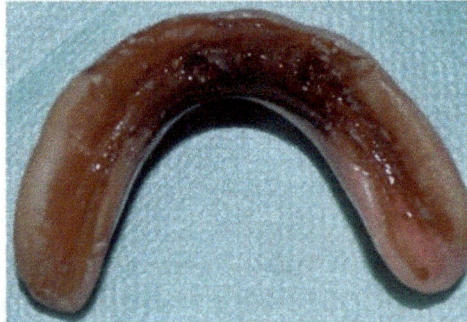

Fig. 2 nicotine staining[74], obtained from sources that are distributed under the Creative Common Attribution license that allows users to use data in any format as long as the credit is given to the author.

Color stability may vary from one material to another. Gulfem concluded that acrylic-based liners are less color stable than silicone-based liners [42]. Hashem proposed that fluorinated soft liners showed improved color stability and better stain resistance [3]. Beverages like tea and coffee can also cause staining of the denture liners. Serra investigated and suggested that coffee can produce more noticeable color changes than tea for the denture soft lining materials [43].

Difficulty in cleaning

Soft liners are readily attacked by Candida albicans, which may lead to denture stomatitis and later require professional prosthetic cleaning [44]. Subsequently, the ultimate drawback of the soft liner is the problem to keep it clean [45].

Several modifications have been made to prevent microbial overgrowth on the surface of a soft liner. Silver has fungicidal and bactericidal properties. Therefore, the composition of the soft liners has been modified by silver nanoparticles in an attempt to improve the material's fungal resistance [46, 47]. Antifungal agents, such as chlorhexidine, fluconazole, and nystatin are also incorporated into a soft liner to provide a slow but continuous release of the drug causing a continued inhibition in the growth of candida Albicans [3].

2. Tissue conditioners

Tissue conditioners (TCs) are a type of reliners that are intended for short term use and formed by mixing polymer in a solution of ethyl alcohol containing plasticizer [37].

2.1 Available products

Following are the commonly available commercial tissue conditioners:

Visco-gel (De Trey), FITT (Kerr), GC Soft-Liner (GC Europe NV), SR-Ivo seal (Ivoclar), Tissue Conditioner (Shofu), and Coe-Comfort (GC America) .

Common problems

Loss of elasticity, water sorption, microbial growth, color change, and inadequate bond retention to base resin are the common problems associated with the use of tissue conditioners. As a result, the materials need to be replaced regularly at short intervals [48].

2.2   Composition

Table 1: Composition of Tissue Conditioners [49]

Powder		Chemical Formula	Function
Polymer beads	Polyethyl methacrylate (PEMA)	$C_6H_{10}O_2)_n$	polymerization
Liquid			
Solvent	Ethyl alcohol (7.5-40.5%)		carrier for plasticizer
Plasticizer	Butylphthalyl butyl glycolate		adds softness and flexibility to material

Tissue conditioners are available as liquid powder systems and Table 1; lists the composition of these conditioners. The plasticizers are low molecular weight liquid esters like dibutyl phthalate, butyl phthalyl butyl glycolate, and butyl benzyl phthalate [50] and serves to decrease the glass transition temperature of the polymer, which leads to softening of the polymer [48, 51]. Following the mixing of the polymer powder with liquid results, gel formation occurs that has viscoelastic properties [52] penetration of PEMA particles into the large molecules of the ester-based plasticizer occurs, the alcohol causes swelling of the polymer and accelerates plasticizer penetration that results in gelation time that is clinically acceptable [48, 50]. Certain factors affect the process of gelation the gelation process. These include polymer molecular weight, powder particle size, the ratio of liquid and powder, the plasticizer amount, and the temperature [52]. A decrease in gelation time was noted when polymer molecular weight was increased, a similar result was obtained when powder liquid proportion was increased [48, 50, 51]. Temperature reportedly has been said to increase the gelation process and therefore in the oral cavity the gelation time is less compared to that with room temperature. Ethanol, acting as a softener and present in the liquid plays a crucial part in the gelation process, but, its rapid evaporation in the oral cavity, causes the tissue conditioner to harden with increased porosity, which ultimately results in the loss of elasticity. Therefore it is recommended to change the tissue conditioner lining after 3-5 days [53]. Ethanol in high quantity causes shrinkage of material in the oral cavity [50] while the highest quantity of ethanol is released within the first 12 h [25].

Due to the differences in the composition and properties of different types of tissue conditioners, it is recommended to choose the material that is best suited for that scenario.

2.3   Clinical applications

Primarily used for the restoration of traumatized oral mucosa caused by denture inflammation, recording functional impressions as shown in Fig. 3, as liners to improve the fit of the acrylic dentures, and for prevention of irritation from the denture. They may also be used for the rehabilitation of cancer patients who need obturation. The viscoelasticity and dimensional stability vary depending on the type of tissue conditioner used [54, 55], therefore, a single Tissue conditioner is not able to fulfill all requirements for applications. Ideally, a tissue conditioner should have high elasticity during masticatory load and have viscoelastic properties to distribute forces evenly [51]. Maintaining denture hygiene with tissue conditioner lining is a major problem since liquid cleaning agents and antiseptic cleansers have shown adverse effects on Tissue conditioners' surfaces [56]. It is recommended that the prosthesis with tissue conditioner should be cleaned with a cotton cloth and disinfected in 0.2% solution of chlorhexidine.

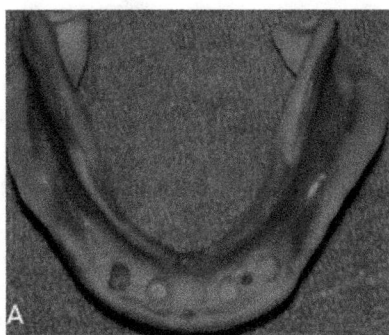

Fig. 3 tissue conditioner used to record functional impression[75], obtained from sources that are distributed under the Creative Common Attribution license that allows users to use data in any format as long as the credit is given to the author.

2.4   Properties

The release of ethanol and plasticizers from tissue conditioners in the mouth is a limitation of tissue conditioner and requires the material to be replaced often [55]. Studies reported a rise in water sorption and solubility within 1 week time period of different tissue conditioners [57, 58]. Another major disadvantage specifically of silicone-based tissue conditioners is the lack of bonding between tissue conditioner liner and the denture plate and is the leading cause of failure of reliners since it creates a suitable growth environment for microbes and formation of plaque and calculus [48, 59].

Adequate bonding between lining material and the denture base is vital for increasing its life in the oral cavity [60]. Improved bond retention can be obtained with roughness in the denture base that can be increased with abrasives and laser treatment [48, 61].

Another important feature of the tissue conditioners is their dimensional properties. With time there is a change due to water sorption and solubility of the conditioner in the mouth. A polymer having greater acceptance for water uptake, there is a volumetric increase in the tissue conditioner while leaching of plasticizers and solvent causes shrinkage in the material, leading to dimensional instability [51]. It is suggested that the time for recording functional impressions should not be more than 24 h. Permaseal and Monopoly, maintain elastic properties for longer durations, increasing its life in the mouth. To achieve these results formulations need to be prepared that have decreased leaching of plasticizers and decreased water sorption [24]. Coating of tissue conditioner surface with Monopoly may prolong the life of the liners up to a year and the smoothness on the surface due to coating also decreases the accumulation of microbial growth.

2.5   Role of tissue conditioner in denture stomatitis

Loss of the plasticizer and roughened surface of tissue conditioner leading to inflammation of the oral mucosa that bears the denture or denture lining surface is a common problem [62].

A study by Okita et al.[63] reported that Coe-Comfort, FITT, Soft-Liner and Visco-gel, show cytotoxicity in vitro, and is said to be greater than that of Polymethyl methacrylate. The release of softening agents from tissue conditioners is a probable disadvantage of these materials. Phthalates that act as softeners in tissue conditioners have reportedly been shown to have estrogenic activity potential[64] new formulations are being tried with the addition of dibutyl citrate or dibutyl sebacate as softeners to overcome the problem of toxicity caused by phthalates [65].

Some of the plasticizers used in tissue conditioners have been shown to retard fungal growth. Leaching of solvent and plasticizers from the surface of tissue conditioners causes it to harden and become porous, hence becoming a good site for microbes to grow. The effectiveness of adding antibacterial and antifungal drugs to TCs has been evaluated. Chow et al [66] study show that when 5% of itraconazole is added to different tissue conditioners, the activity remains the greatest within the first three days, after which reliners should be replaced for better healing.

Studies have shown that surface roughness of dentures and resilient liners harbors microbial growth by increasing the accumulation of microorganisms by enhancing the adhesion of microorganisms and which in turn can lead to candida associated denture stomatitis [67]. A study by Radnai et al. [68] showed that chlorhexidine gluconate gel incorporated in tissue Conditioner did not affect the candida Albicans growth, while the addition of miconazole showed a dose-related inhibitory effect on candida growth.

Applications of Polymers in Surgery   Materials Research Forum LLC
Materials Research Foundations **123** (2022) 175-194   https://doi.org/10.21741/9781644901892-7

It has been suggested for further studies silver zeolite be added to tissue conditioners since it has reportedly shown to have a dose dependent inhibition on Candida albicans.

3.   Hard reliners

Hard relines are non-resilient materials that contain an acrylic resin that is similar in composition to the acrylic used in the fabrication of denture bases. These materials are mainly used in chairside relining of the dentures in dental clinics. They are less comfortable but are famous for their durability, stability in the oral environment, and easy cleaning [69].

3.1   Composition

Hard reliners are composed are available as powder liquids systems, the powder is composed of Polymethyl methacrylate with Benzoyl peroxide as initiator, and liquid is composed of methacrylate monomers with tertiary amine as chemical activator. The composition is listed in Table 2, following the mixing of polymer powder with monomer, the resin readily undergoes polymerization at room/ mouth temperatures [71].

Table 2: Composition of Hard reliners

Type 1 :		
Powder:	**Chemical formula**	**Function:**
Polymethylmethacrylate	$(C_5O_2H_8)_n$	Polymer beads
Benzoyl peroxide		Initiator
Liquid:		
Methylmethacrylate	$C_5H_8O_2$	Monomer
Di-n-butyl phthalate		Plasticizer
Tertiary amine		Chemical Activator
Type 2		
Powder :		
Polymethylmethacrylate		Polymer beads
Benzoyl peroxide		Initiator
Liquid :		
Butyl methacrylate	$C_8H_{14}O_2$	Monomer
Di-methacrylate		Cross linking agent
Tertiary amine		Chemical Activator
Pigments		

3.2 Properties

Viscoelasticity enables the lining material to absorb and evenly distribute the masticatory forces. Sheridan in 2018, evaluated the viscoelastic properties of hard denture lining materials using viscoelastometer. It was concluded that the hard reline transmitted less stress to the underlying mucosa and alveolar bone. Therefore, these materials have far better able to absorb energy and relieve stresses under masticatory force [70].

The lining materials applied must form a strong bond with the denture base to ensure the proper functioning of the denture. In 2014, Mayank Lau tested the shear and tensile strength of the hard relines and found that the strength values of the hard relines were significantly high making the material more durable [71]. Therefore, it was concluded that hard relines can be used safely in high stress bearing areas.

3.3 Biocompatibility

Due to direct contact of hard reliner with oral soft tissues. These materials should not be toxic or irritating. In 2012, Ayse Atay assessed the cytotoxicity of the hard denture relines at different incubation periods. The results concluded that all the test materials showed good biocompatibility [72]. Hard relines are the least cytotoxic and safe to use in clinical practice.

3.4 Advantages and disadvantages

Clinically the hard reliners are more durable, have better retention, improved adhesion with the denture base resin and the process is less time consuming. Their low glass transition temperature, dimensional instability and irritation due to direct contact with mucosa are its disadvantages.

3.5 Clinical limitations

Hard relines are made with a harder plastic therefore clinically, they are not preferred [73]. Also, they are not recommended in patients with sensitive gums. Unlike soft reline, these materials fail to provide the cushioning effect, therefore their use is also contraindicated in thin mucosa or resorbed ridges [70]. The methacrylate monomer is a known irritant that may sensitize the oral tissues after direct placement of the hard relines [38]. Another problem associated with the hard reline is the increased thickness of the palate for upper dentures. This increase in thickness is often unpleasant for the patients [38].
Due to their limitations, the use of hard reliners is decreasing with time.

Applications of Polymers in Surgery   Materials Research Forum LLC
Materials Research Foundations **123** (2022) 175-194   https://doi.org/10.21741/9781644901892-7

Conclusion

The function of polymeric soft denture lining materials in the prevention and management of chronic tissue irritation from dentures is an exceptional way to reduce patient discomfort and is beneficial for preserving the health of the residual denture supporting tissue. However, the limitations of these materials mainly including lack of adhesion between acrylic base and soft liners, enhancing microbial growth, and leaching of plasticizers making them less durable. Tissue conditioners that majorly act as cushioning material are intended only for short term use and lose their advantageous properties when used for longer durations and cause irritation to the mucosa, denture plaque accumulation leading to worsening of denture stomatitis. Therefore, the tissue conditioner liners should be replaced within a specified period of 3-5 days. Since the limitations outweigh the advantages of hardliners their use is diminishing with time. There's a hope that broader applications will establish in the future once the current drawbacks of the available commercial materials are dealt with.

References

[1] The Glossary of Prosthodontic Terms: Ninth Edition, J Prosthet Dent. 117(5s) (2017) e1-e105. https://doi.org/10.1016/j.prosdent.2016.12.001

[2] L.T. Garcia, J.D. Jones, Soft liners, Dent. Clin. N. A. 48(3) (2004) 709-20, vii. https://doi.org/10.1016/j.cden.2004.03.001

[3] M.I. Hashem, Advances in Soft Denture Liners: An Update, J. Contemp. Dent. Pract. 16(4) (2015) 314-8. https://doi.org/10.5005/jp-journals-10024-1682

[4] A. Nowakowska-Toporowska, Z. Raszewski, W. Wieckiewicz, Color change of soft silicone relining materials after storage in artificial saliva, J Prosthet Dent. 115(3) (2016) 377-380. https://doi.org/10.1016/j.prosdent.2015.08.022

[5] A. Naeem, Int. Multidiscip. Res. J. www.allsubjectjournal.com, 2015.

[6] H.-S. Cha, B. Yu, Y.-K. Lee, Changes in stress relaxation property and softness of soft denture lining materials after cyclic loading, Dent. Mater. J. 27(3) (2011) 291-297. https://doi.org/10.1016/j.dental.2010.11.004

[7] A.M. Dimiou, K. Michalakis, A. Pissiotis, Influence of thickness increase of intraoral autopolymerizing hard denture base liners on the temperature rise during the polymerization process, J Prosthet Dent. 111(6) (2014) 512-520. https://doi.org/10.1016/j.prosdent.2013.07.021

[8] Y. Tanimoto, H. Saeki, S. Kimoto, T. Nishiwaki, N. Nishiyama, Evaluation of adhesive properties of three resilient denture liners by the modified peel test method, Acta Biomater. 5(2) (2009) 764-9. https://doi.org/10.1016/j.actbio.2008.08.021

[9] W.J. O'Brien, quintpub, www.quintpub.com, 2002.

[10] H. Murata, T. Hamada, S. Sadamori, Relationship between viscoelastic properties of soft denture liners and clinical efficacy, Jpn Dent Sci Rev. 44 (2008). https://doi.org/10.1016/j.jdsr.2008.06.001

[11] D.K. Ayappa, jpsr, https://www.jpsr.pharmainfo.in, 2019.

[12] A. Harrison, Temporary soft lining materials. A review of their uses, Br. Dent. J. 151(12) (1981) 419-22. https://doi.org/10.1038/sj.bdj.4804726

[13] S. Kreve, A.C. Dos Reis, Denture Liners: A Systematic Review Relative to Adhesion and Mechanical Properties, Sci. World J. (2019) 6913080. https://doi.org/10.1155/2019/6913080

[14] B.T. Léon, A.A.D.B. Cury, R. Garcia, Loss of residual monomer from resilient lining materials processed by different methods, 2008.

[15] T. Ohkubo, M. Oizumi, T. Kobayashi, Influence of methylmercaptan on the bonding strength of autopolymerizing reline resins to a heat-polymerized denture base resin, Dent. Mater. J. 28(4) (2009) 426-432. https://doi.org/10.4012/dmj.28.426

[16] W. Santawisuk, W. Kanchanavasita, C. Sirisinha, C. Harnirattisai, Dynamic viscoelastic properties of experimental silicone soft lining materials, Dent. Mater. J. (2010) 1007120040-1007120040. https://doi.org/10.4012/dmj.2009-126

[17] J.M.F.K. Takahashi, R.L.X. Consani, G.E.P. Henriques, M.A. de Arruda Nóbilo, M.F. Mesquita, Effect of Accelerated Aging on Permanent Deformation and Tensile Bond Strength of Autopolymerizing Soft Denture Liners, J Prosthodont. 20(3) (2011) 200-204. https://doi.org/10.1111/j.1532-849X.2010.00679.x

[18] E.B. Moffa, E.T. Giampaolo, F.E. Izumida, A.C. Pavarina, A.L. Machado, C.E. Vergani, Colour stability of relined dentures after chemical disinfection. A randomised clinical trial, J. Dent. 39 (2011) e65-e71. https://doi.org/10.1016/j.jdent.2011.10.008

[19] F.E. Izumida, J.H. Jorge, R.C. Ribeiro, A.C. Pavarina, E.B. Moffa, E.T. Giampaolo, Surface roughness and Candida albicans biofilm formation on a reline resin after long-term chemical disinfection and toothbrushing, J Prosthet Dent. 112(6) (2014) 1523-1529. https://doi.org/10.1016/j.prosdent.2014.06.001

[20] F. Valentini, M.S. Luz, N. Boscato, T. Pereira-Cenci, Surface Roughness Changes in Denture Liners in Denture Stomatitis Patients, Int J Prosthodont. 30(6) (2017) 561-564. https://doi.org/10.11607/ijp.5108

[21] H.J. Raval, N. Mahajan, Y.G. Naveen, R. Sethuraman, A Three Month Comparative Evaluation of the Effect of Different Surface Treatment Agents on the Surface Integrity and Softness of Acrylic based Soft Liner: An In vivo Study, J. Clin. Diagnostic Res. 11(9) (2017) Zc88-zc91. https://doi.org/10.7860/JCDR/2017/28604.10680

[22] A. Ogawa, S. Kimoto, H. Saeki, M. Ono, N. Furuse, Y. Kawai, The influence of patient characteristics on acrylic-based resilient denture liners embedded in maxillary complete dentures, J. Prosthodont. Res. 60(3) (2016) 199-205. https://doi.org/10.1016/j.jpor.2015.12.001

[23] K. Singh, P. Chand, B.P. Singh, C.B. Patel, Study of the effect of surface treatment on the long term effectiveness of tissue conditioner, J. Oral Sci. 52(2) (2010) 261-5. https://doi.org/10.2334/josnusd.52.261

[24] H.S. Malmström, N. Mehta, R. Sanchez, M.E. Moss, The effect of two different coatings on the surface integrity and softness of a tissue conditioner, J Prosthet Dent. 87(2) (2002) 153-7. https://doi.org/10.1067/mpr.2002.121407

[25] K.B. Shylesh, R.B. Hallikerimath, V.V. Rajeev, S. Ganesh, S. Meenakshi, Evaluation of Physical Properties of Tissue Conditioning Materials as used in Functional Impression - A Lab Study, J Int Oral Health. 5(3) (2013) 20-7.

[26] R. Osathananda, C. Wiwatwarrapan, Surface treatment with methyl formate-methyl acetate increased the shear bond strength between reline resins and denture base resin, Gerodontology 33(2) (2016) 147-54. https://doi.org/10.1111/ger.12120

[27] T. Maeda, G. Hong, S. Sadamori, T. Hamada, Y. Akagawa, Durability of peel bond of resilient denture liners to acrylic denture base resin, J. Prosthodont. Res. 56(2) (2012) 136-41. https://doi.org/10.1016/j.jpor.2011.05.001

[28] G. Muralidhar, C.L. Satish Babu, S. Shetty, Integrity of the interface between denture base and soft liner: a scanning electron microscopic study, J. Indian Prosthodont. Soc. 12(2) (2012) 72-7. https://doi.org/10.1007/s13191-011-0111-8

[29] L.V. Lassila, M.M. Mutluay, A. Tezvergil-Mutluay, P.K. Vallittu, Bond strength of soft liners to fiber-reinforced denture-base resin, J Prosthodont. 19(8) (2010) 620-4. https://doi.org/10.1111/j.1532-849X.2010.00642.x

[30] J.H. Kim, H.C. Choe, M.K. Son, Evaluation of adhesion of reline resins to the thermoplastic denture base resin for non-metal clasp denture, Dent Mater J. 33(1) (2014) 32-8. https://doi.org/10.4012/dmj.2013-121

[31] B.J. Kim, H.S. Yang, M.G. Chun, Y.J. Park, Shore hardness and tensile bond strength of long-term soft denture lining materials, J Prosthet Dent. 112(5) (2014) 1289-97. https://doi.org/10.1016/j.prosdent.2014.04.018

[32] Y.W. Cavalcanti, M.M. Bertolini, A.A. Cury, W.J. da Silva, The effect of poly(methyl methacrylate) surface treatments on the adhesion of silicone-based resilient denture liners, J Prosthet Dent. 112(6) (2014) 1539-44. https://doi.org/10.1016/j.prosdent.2013.07.031

[33] A. Usumez, O. Inan, F. Aykent, Bond strength of a silicone lining material to alumina-abraded and lased denture resin, J. Biomed. Mater. Res. 71(1) (2004) 196-200. https://doi.org/10.1002/jbm.b.30078

[34] M. Kazanji, A. Al-Ali, Z. Ahmad, Water absorption and solubility of coated and non-coated silicone-based permanent soft liner, (2010).

[35] I. Hayakawa, N. Akiba, E. Keh, Y. Kasuga, Physical properties of a new denture lining material containing a fluoroalkyl methacrylate polymer, J Prosthet Dent. 96(1) (2006) 53-58. https://doi.org/10.1016/j.prosdent.2006.05.012

[36] M.X. Pisani, V.M.F. Leite, M.M. Badaró, A.d.L. Malheiros-Segundo, H.d.F.d.O. Paranhos, C.H.L.d. Silva, Soft denture liners and sodium perborate: sorption, solubility and color change, Braz. J. Oral Sci. 14 (2015) 219-223. https://doi.org/10.1590/1677-3225v14n3a09

[37] B. Dorocka-Bobkowska, D. Medynski, M. Prylinski, Recent advances in tissue conditioners for prosthetic treatment: A review, Adv Clin Exp Med. 26(4) (2017) 723-728. https://doi.org/10.17219/acem/62634

[38] M.G. Waters, R.G. Jagger, R.W. Winter, Water absorption of (RTV) silicone denture soft lining material, J Dent. 24(1-2) (1996) 105-8. https://doi.org/10.1016/0300-5712(95)00032-1

[39] Y. Kasuga, N. Akiba, S. Minakuchi, T. Uchida, N. Matsushita, M. Hishimoto, I. Hayakawa, Development of soft denture lining materials containing fluorinated monomers, Nihon Hotetsu Shika Gakkai zasshi 52(2) (2008) 183-8. https://doi.org/10.2186/jjps.52.183

[40] S. Sudhapalli, Time Dependent Effect of a Denture Cleanser on the Sorption and Solubility of Four Soft Liners-An Invitro Study, J. clin. diagn. res. (2016) ZC100-ZC103. https://doi.org/10.7860/JCDR/2016/16952.7682

[41] P. Imirzalioglu, O. Karacaer, B. Yilmaz, I. Ozmen Msc, Color stability of denture acrylic resins and a soft lining material against tea, coffee, and nicotine, J. Prosthodont. 19(2) (2010) 118-24. https://doi.org/10.1111/j.1532-849X.2009.00535.x

[42] G. Ergun, I.C. Nagas, Color stability of silicone or acrylic denture liners: an in vitro investigation, Eur J Dent 1(3) (2007) 144-151. https://doi.org/10.1055/s-0039-1698330

[43] S. Oguz, M.M. Mutluay, O.M. Dogan, B. Bek, uuml, lent, Color Change Evaluation of Denture Soft Lining Materials in Coffee and Tea, Dent. Mater. J. 26(2) (2007) 209-216. https://doi.org/10.4012/dmj.26.209

[44] H. Nikawa, T. Yamamoto, T. Hamada, Effect of components of resilient denture-lining materials on the growth, acid production and colonization of Candida albicans, J. Oral. Rehabil. 22(11) (1995) 817-24. https://doi.org/10.1111/j.1365-2842.1995.tb00228.x

[45] T. Baygar, A. Ugur, N. Sarac, U. Balci, G. Ergun, Functional denture soft liner with antimicrobial and antibiofilm properties, J. Dent. Sci. 13(3) (2018) 213-219. https://doi.org/10.1016/j.jds.2017.10.002

[46] C. Fan, L. Chu, H.R. Rawls, B.K. Norling, H.L. Cardenas, K. Whang, Development of an antimicrobial resin--a pilot study, Dent. Mater. J. 27(4) (2011) 322-8. https://doi.org/10.1016/j.dental.2010.11.008

[47] G. Chladek, I. Barszczewska-Rybarek, J. Lukaszczyk, Developing the procedure of modifying the denture soft liner by silver nanoparticles, Acta Bioeng Biomech. 14(1) (2012) 23-9. https://doi.org/10.3390/ijms14010563

[48] S. Rodrigues, V. Shenoy, T. Shetty, Resilient liners: a review, J. Indian Prosthodont. Soc. 13(3) (2013) 155-64. https://doi.org/10.1007/s13191-012-0143-8

[49] A.W.G.W. John F. McCabe, Applied Dental Materials Ninth Edition, (2008).

[50] M.B. S. Parker, The effect of particle size on the gelation of tissue conditioners, Biomaterials 22 (2001) 2039-2042. https://doi.org/10.1016/S0142-9612(00)00391-4

[51] M.I. Hashem, Advances in Soft Denture Liner: An Update, J Contemp Dent Pract. 16(4) (2015) 314-318. https://doi.org/10.5005/jp-journals-10024-1682

[52] H.S. Abbas Monzavi, Mohammad Atai, Marzieh Alikhasi, Vahieh Nazzari, Sadigheh Shiekhzadeh, Comparative Evaluation of Physical Properties of Four Tissue Conditioners Relined To Modeling Plastic Material, J. Dent. 10(6) (2013) 506-515.

[53] T.H. H. MURATA, HARSHINI, K. TOKI & H. NIKAWA, Effect of addition of ethyl alcohol on gelation and viscoelasticity of tissue conditioner, J. Oral Rehabil. 28 (2008) 48-54. https://doi.org/10.1046/j.1365-2842.2001.00638.x

[54] H. Murata, M. Kawamura, T. Hamada, S. Saleh, U. Kresnoadi, K. Toki, Dimensional stability and weight changes of tissue conditioners, J. Oral Rehabil. 28(10) (2001) 918-23. https://doi.org/10.1046/j.1365-2842.2001.00736.x

[55] H. Murata, T. Hamada, E. Djulaeha, H. Nikawa, Rheology of tissue conditioners, J Prosthet Dent . 79(2) (1998) 188-99. https://doi.org/10.1016/S0022-3913(98)70215-X

[56] N.L. Jacobsen, D.L. Mitchell, D.L. Johnson, R.A. Holt, Lased and sandblasted denture base surface preparations affecting resilient liner bonding, J Prosthet Dent . 78(2) (1997) 153-8. https://doi.org/10.1016/S0022-3913(97)70119-7

[57] D.W. Jones, E.J. Sutow, B.S. Graham, E.L. Milne, D.E. Johnston, Influence of Plasticizer on Soft Polymer Gelation, J. Dent. Res. 65(5) (1986) 634-642. https://doi.org/10.1177/00220345860650050101

[58] F.N. Sampaio, J.R. Pinto, C.P. Turssi, R.T. Basting, Effect of sealant application and thermal cycling on bond strength of tissue conditioners to acrylic resin, Braz. Dent. J. 24(3) (2013) 247-52. https://doi.org/10.1590/0103-6440201302118

[59] W. Więckiewicz, J. Kasperski, M. Więckiewicz, M. Miernik, W. Król, The adhesion of modern soft relining materials to acrylic dentures, Adv Clin Exp Med. 23(4) (2014) 621-5. https://doi.org/10.17219/acem/37242

[60] G.L. Polyzois, M.J. Frangou, Influence of curing method, sealer, and water storage on the hardness of a soft lining material over time, J Prosthodont. 10(1) (2001) 42-5. https://doi.org/10.1053/jpro.2001.23546

[61] C.A. Chaves, A.L. Machado, C.E. Vergani, R.F. de Souza, E.T. Giampaolo, Cytotoxicity of denture base and hard chairside reline materials: a systematic review, J Prosthet Dent. 107(2) (2012) 114-27. https://doi.org/10.1016/S0022-3913(12)60037-7

[62] N. Chopde, A. Pharande, M.N. Khade, Y.R. Khadtare, S.S. Shah, A. Apratim, In vitro antifungal activity of two tissue conditioners combined with nystatin, miconazole and fluconazole against Candida albicans, J. Contemp. Dent. Pract. 13(5) (2012) 695-8. https://doi.org/10.5005/jp-journals-10024-1211

[63] N. Okita, A. Hensten-Pettersen, In vitro cytotoxicity of tissue conditioners, J Prosthet Dent. 66(5) (1991) 656-9. https://doi.org/10.1016/0022-3913(91)90448-6

[64] H. Murata, H. Chimori, T. Hamada, J.F. McCabe, Viscoelasticity of dental tissue conditioners during the sol-gel transition, J Dent Res. 84(4) (2005) 376-81. https://doi.org/10.1177/154405910508400416

[65] M. Nishijima, Y. Hashimoto, M. Nakamura, Cytocompatibility of new phthalate ester-free tissue conditioners in vitro, Dent Mater J. 21(2) (2002) 118-32. https://doi.org/10.4012/dmj.21.118

[66] C.K. Chow, D.W. Matear, H.P. Lawrence, Efficacy of antifungal agents in tissue conditioners in treating candidiasis, Gerodontology. 16(2) (1999) 110-8. https://doi.org/10.1111/j.1741-2358.1999.00110.x

[67] M.E. Brosky, I.J. Pesun, B. Morrison, J.S. Hodges, J.H. Lai, W. Liljemark, Clinical evaluation of resilient denture liners. Part 2: Candida count and speciation, J Prosthodont. 12(3) (2003) 162-7. https://doi.org/10.1016/S1059-941X(03)00041-X

[68] M. Radnai, R. Whiley, T. Friel, P.S. Wright, Effect of antifungal gels incorporated into a tissue conditioning material on the growth of Candida albicans, Gerodontology. 27(4) (2010) 292-6. https://doi.org/10.1111/j.1741-2358.2009.00337.x

[69] woodside denture centre, What are Denture Relines?, 2018, Oct 1. https://woodsidedenturecentre.com/what-are-denture-relines/.

[70] S. Eissa, D. Zaki, A. Elboraey, A.R. Moussa, Viscoelastic properties of hard and soft denture base reline materials, Int. J. Adv. (2018).

[71] M. Lau, G.S. Amarnath, B.C. Muddugangadhar, M.U. Swetha, K.A.A.K. Das, Tensile and shear bond strength of hard and soft denture relining materials to the conventional heat cured acrylic denture base resin: An In-vitro study, J Int Oral Health . 6(2) (2014) 55-61.

[72] A. Atay, V. Bozok Çetintaş, E. Cal, B. Kosova, A. Kesercioglu, P. Guneri, Cytotoxicity of hard and soft denture lining materials, Dent. Mater. J. 31 (2012) 1082-6. https://doi.org/10.4012/dmj.2012-209

[73] d.a.i. br, What to Know About Denture Relines, 2019, Jun 3. https://www.denturesandimplantsbr.com/what-to-know-about-denture-relines.

[74] G. Chladek, J. Zmudzki, J. Kasperski, Long-Term Soft Denture Lining Materials, Materials .7(8) (2014) 5816-5842. https://doi.org/10.3390/ma7085816

[75] B.-K. Lee, S.-H. Park, C.-H. Lee, J.-H. Cho, Implant overdenture impressions using a dynamic impression concept, J Adv Prosthodont. 6(1) (2014) 66. https://doi.org/10.4047/jap.2014.6.1.66

Applications of Polymers in Surgery
Materials Research Foundations **123** (2022) 195-215

Materials Research Forum LLC
https://doi.org/10.21741/9781644901892-8

Chapter 8

Polymers in Guided Bone Regeneration

Rafaella Moreno Barros[1], Demis Ferreira de Melo[1], Bruna Galdorfini Chiari-Andréo[2], João Augusto Oshiro-Júnior[1]*

[1] Pharmaceutical Sciences Postgraduate Program, Center for Biological and Health Sciences, State University of Paraíba - UEPB, Campina Grande, Paraíba, Brazil

[2] Department of Biological and Health Sciences, University of Araraquara – UNIARA, Araraquara, São Paulo, Brazil

* joaooshiro@yahoo.com.br

Abstract

In dental debilitating conditions, such as edentulism, the patient must have sufficient alveolar crest bone mass to guarantee the possibility of corrective interventions, such as implants. The decrease in bone tissue is caused by the body's natural reabsorption, which begins after tooth loss. Some surgical techniques could be used to solve this problem, such as guided bone regeneration. In this technique, a membrane developed with biomaterials is used, which aims to act as physical barrier to prevent the appearance of soft tissue and maintain the bone defect space, ensuring maximum regeneration. This membrane must be biocompatible, have specific rigidity to maintain the space, prevent the migration of epithelial cells and ensure the resorption time after bone tissue regeneration. This chapter will address the polymeric materials most used in the development of membranes for guided bone regeneration process, addressing their physicochemical characteristics, advantages, disadvantages, among other important aspects.

Keywords

Bone Tissue, Regeneration, Biocompatible Membrane, Implants, Edentulism

Contents

1.   Introduction

Edentulism is characterized by the total or partial loss of teeth, being an irreversible and debilitating condition. The World Health Organization (WHO) considers it a serious public health problem and recommends the adoption of the theme in public health policies. It is a multifactorial process and its origin can be congenital or acquired (caries, trauma, periodontitis, oral cancer, population aging, socioeconomic factors, culture, among others) [1].

Currently, evaluating the rate of edentulism in the population, data indicates that women is mainly affect and, also, that the number of cases of this condition is decreasing in developed countries, on the other hand, it is growing in developing countries [2,3].

Among the consequences, the patient may have difficulty of chewing, swallowing and speaking, in addition to the aesthetic factors that can lead, for example, to loss of self-esteem. Such functions (phonetic and aesthetic) can be corrected through prosthodontic rehabilitation that will replace the missing teeth. When evaluating the case, the dentist will indicate the best treatment, showing the risks, benefits and costs of the modality, as socioeconomic status is one of the factors that hinder the search for treatment [4].

Depending on the severity, the following may be indicated by the professional: fixed partial dentures, removable partial dentures, overdentures or implants. Implants is a procedure considered to be of high cost, which makes the access difficult. However, the worldwide implant market moved around US$ 9.1 billion in 2018 and US$ 13 billion in 2020, showing a trend of growth demand for the procedure. Consequently, it drives the development of new innovative, safe and low-cost materials for dentistry implants [5,6].

For the implant surgery to be successful, the patient must have sufficient alveolar crest bone mass to guarantee the fixation. This decrease in bone tissue is caused by the body's natural reabsorption, which begins after tooth loss. Onlay and inlay grafts, crest expansion, distraction osteogenesis and guided bone regeneration (GBR) are some surgical techniques used to solve this problem [7].

Among these techniques, GBR stands out for promoting tissue restoration through the process of cell selectivity. A membrane developed with biomaterials is used, which aims to act as physical barrier to prevent the appearance of soft tissue and maintain the bone defect space, ensuring maximum regeneration. For this, it is necessary to comply with the cellular steps of osteoinduction, osteoconduction and osteogenesis, separately or in synergy. Therefore, this membrane must be biocompatible, have specific rigidity to maintain the space, prevent the migration of epithelial cells and ensure the resorption time after bone tissue regeneration [8].

Considering the relevance of this theme, this chapter will address the polymeric materials most used in the development of membranes for GBR process, addressing their physicochemical characteristics, advantages, disadvantages, among other important aspects.

2. Polymers used in non-resorbable membranes

Considering the relevance of this theme, this chapter will address the polymeric materials most used in the development of membranes for GBR process, addressing their physicochemical characteristics, advantages, disadvantages, among other important aspects.

Non-resorbable membranes have several properties that justify their extensive use in GBR, such as biocompatibility, ability to maintain the bone defect space for longer than non-resorbable membranes, easy handling in clinical practice, in addition to having more predictable regeneration profiles due to its adequate mechanical strength [9,10]. However, one of the main disadvantages of non-resorbable membranes is the reduced adhesiveness to the cells and the lack of connection capacity with the bone tissue to produce osseointegration, which reflects the need for a second surgical procedure, 4 to 6 weeks later of implantation, to remove the membrane and with the possibility of compromising the tissue that was regenerated during this procedure [9,11].

Polytetrafluoroethylene (PTFE) (Fig. 1) is the main representative of polymers used in the development of non-resorbable membranes used in GBR and can be found in expanded (e-PTFE), high density (d-PTFE) and titanium reinforced (Ti-PTFE) forms [12,13].

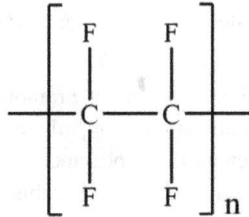

Figure 1.   Chemical structure of PTFE.

2.1   Polytetrafluoroethylene

Polytetrafluoroethylene (PTFE), formed by the polymerization of tetrafluoroethylene (TFE), is considered the gold standard for GBR and has remarkable characteristics such as excellent chemical and thermal stability, high fracture resistance and low surface friction [14]. These characteristics can be explained by the high degree of crystallinity, the strong carbon-carbon and carbon-fluorine bonds of the PTFE structure and the size of the fluorine atoms, which promote a uniform and continuous coverage of the carbon-carbon bonds, protecting them of chemical attacks [15,16]. The first applications of PTFE in GBR were described by Dahlin et al. in 1988, who used PTFE membranes to isolate bone defects in rat jaws from connective tissue [17]. Based on its structure, PTFE can be divided into expanded-PTFE (e-PTFE) and high-density PTFE (d-PTFE).

e-PTFE membranes, commercially available as GORE® TEX® (W.L. Gore & Associates Inc., USA), have two different microstructures, the coronal edge and the occlusive portion, which create a porous environment. The first has an open microstructure, with an intermodal distance (space formed between the nodes during a thermomechanical stretch) between 25-30 μm, which facilitates the fixation of the collagen fibers to stabilize the membrane until it becomes fixed in place, with stabilization of the host-tissue interface, in addition to favoring the formation of initial clots. The second microstructure, in turn, has an intermodal distance less than 8 μm and, therefore, allows the influx of nutrients while preventing the infiltration of cells from other tissues, such as epithelial cells, for example [18,19]. In the pioneering study of the use of PTFE in GBR in bone defects in rat jaws, Dahlin et al. [17] reported greater bone growth in the group that used e-PTFE membranes (3.8 mm) compared to the untreated control group (2.2 mm), in addition to the absence of soft tissue cells in the group treated by the membranes, which evidenced their barrier function.

However, the porous environment formed by these two layers makes e-PTFE membranes susceptible to microbial contamination, especially in the oral cavity, which can lead to infections, which can compromise bone growth and osseointegration [20,21]. As an alternative to overcome these problems, d-PTFE membranes which pores with a size of approximately 0.2 µm were developed. Due to their reduced pore size, bacterial infiltration in the bone development area is eliminated. Furthermore, this smaller pore size compared to e-PTFE membranes eliminates the requirement for primary closure of the soft tissue [19,22]. These membranes are commercially available under trade names of High-density GORE® TEX® (W.L. Gore & Associates Inc., EUA), Cytoplast® (Osteogenics Biomedical, EUA), TefGen FD® (Lifecore Biomedical Inc., EUA) e Non-resorbable ACE (Surgical supply Inc., EUA).

Both PTFE membranes can have their properties improved with the use of titanium as reinforcement (Ti-PTFE), commercially available as GORE® TEX® TI (W.L. Gore & Associates Inc., EUA) [9,23]. This metal has good biocompatibility, mechanical resistance, low density, durability and corrosion resistance, in addition to having excellent osseointegration by direct connection with bone through the extracellular matrix [24]. The reinforcement of PTFE membranes with titanium increases their rigidity, which guarantees the maintenance of the space needed for bone tissue cells migrate to this site, in addition to promoting reduced mucosal compression during the regeneration period [25–27].

3. Polymers used in absorbable membranes

Absorbable membranes are alternatives to non-resorbable membranes and must comply with several properties to be considered ideal, such as biocompatibility, not inducing inflammatory reactions, being fully reabsorbable, being easy to handle, safely exclude non-bone tissue from the defect, resistant to microbial infections, in addition to have predictable resorption time with bone formation [28]. However, commercially available absorbable membranes cannot maintain an adequate space between the defect and the soft tissue, unless the morphology of the defect is favorable to the formation of this space [29]. They also have unpredictable time and degree of degradation, which directly affects bone formation [9]. Furthermore, the products generated during the degradation of these membranes can interfere with the mechanical stability of the membrane and induce inflammatory reactions [10,30].

Despite these disadvantages of these membranes, they have advantages that justify their use in GBR. The main one is that there is no need for a second surgical procedure to remove the membrane, as occurs in non-resorbables, which simplifies the procedure and makes it less expensive, in addition to avoiding tissue damage and the risk of additional morbidities [31]. Among non-resorbable membranes available on the market, the vast majority are composed of synthetic aliphatic polyesters, such as polylactic acid, polyglycolic acid, ε-polycaprolactone and its copolymers (eg polylactic co-glycolic acid), in addition to polyurethanes and natural polymers, like chitosan [32].

3.1   Aliphatic polyesters

Aliphatic polyesters are the main representatives of synthetic biodegradable polymers used in the biomedical field. The most common synthesis route for this class of polymers is through the polymerization process via ring opening (ROP) from some cyclic lactone that occur in nature, such as lactide, glycolide and ε-caprolactone [33,34], which leads to the formation of polylactic acid (PLA), polyglycolic acid (PGA) and ε-polycaprolactone (PCL) (Fig. 2), respectively. In addition to promoting polymerization, this process favors the biodegradation of these polymers formed by reactions catalyzed by hydrolytic enzymes (eg lipase, elastase), since the opening of the rings by ROP causes an increase in molecular flexibility and improvement in the adjustment of the polymer chain in the enzymatic active site, in relation to cyclic esters [35].

The first applications of aliphatic polyesters in the biomedical area were in the production of absorbable sutures in the 60s [36]. At GBR, this class of synthetic polymers is attractive for its ease of industrial processing, adjustment of mechanical properties and biodegradability of membranes by polymer composition, in addition to the possibility of loading with bioactive compounds that can improve GBR [24,37–39].

Among the aliphatic polymers, PLA is used in the preparation of resorbable membranes due to thermal plasticity, adequate mechanical properties and biocompatibility [40–42]. Among commercially available PLA membranes, Epi-Guide® (Kensey Nesh Corporation, EUA), it is capable of maintaining its structure and functions for a period of five months and is completely reabsorbed after one year [43]. The layer that remains in contact with the bone has limited porosity that favors fluid absorption and membrane adhesion, in addition to preventing fibroblast infiltration into the bone defect. The layer in contact with the membrane, in turn, has greater porosity and allows the infiltration and fixation of fibroblasts [40].

A

B

C

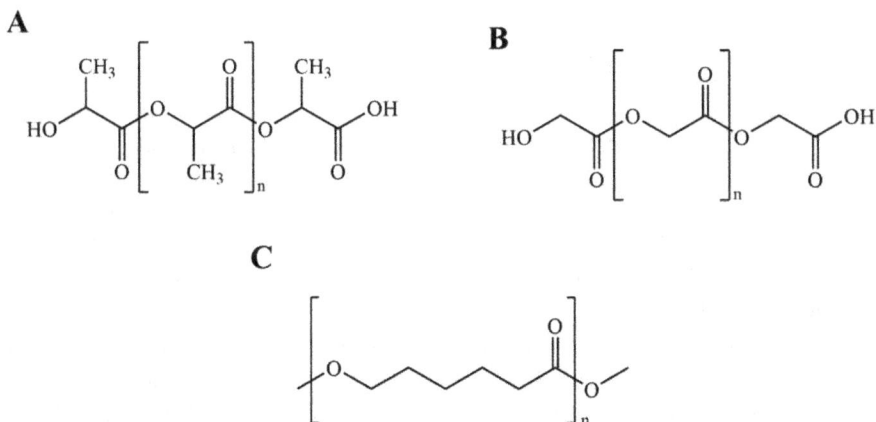

*Figure 2.   Chemical structures of polylactic acid (PLA - A), polyglycolic acid (PGA -
B) and ε-polycaprolactone (PCL - C).*

Another membrane composed of this commercially available polymer is Atrisorb®
(Tolmar Inc., USA). It consists of PLA dissolved in N-methyl-2-pyrrolidone and, as it is
presented in the liquid form, the membrane can be molded according to the morphology of
the bone defect. Complete resorption of this membrane was observed between 6-12 months
after implantation in periodontal defects [44]. The N-methyl-2-pyrrolidone used in this
membrane, in addition to solubilizing PLA, has the ability to inhibit osteoclast activity and
increase bone formation, via increased bioactivity and bioavailability of autogenous bone
morphogenetic proteins (BMP), which results in preservation of existing bone and
enhancement of BMPs, as described by Schneider et al. [45].

In order to modulate the membrane degradation time, polyglycolic acid (PGA) can be used
in substitution or copolymerized with PLA, as it has greater hydrophilicity compared to
PLA and greater susceptibility to enzymatic degradation. This fact can be explained by the
absence of asymmetric methyl groups in PGA monomers compared to PLA, which reduces
lipophilicity and the steric hindrance effect for enzymatic hydrolysis. From these
properties, the PGA has resorption onset between 1-2 months, with complete resorption
between 6-12 months [46–48]. Among the commercial options for PGA membranes,
Biofix® (Biocon, Finland) is a partially fibrous asymmetric membrane that has a resorption
time between 6-12 months and promotes isolation of the bone defect against gingival and
bacterial cells [9,49,50].

Between the membranes composed of copolymers of PLA and PGA (PLGA), Vicryl (Johnson & Johnson, USA), a tissue membrane developed by fibers of PLA and PGA in a 1:9 (m/m) ratio, called Polyglactin 910, is inert, non-immunogenic and has the ability to preserve its mechanical properties between 3-4 weeks after implantation [32,40]. Resolut Adapt® e Resolut Adapt® LT (W.L. Gore & Associates Inc., EUA) are other membranes composed of PLGA that exhibit a barrier time between 8-10 and 16-24 weeks, respectively, and are completely resorbable in the period between 5-6 months, which is about 50% of the average resorption time of PLA membranes [43]. Wadhawan et al. [51] compared the effectiveness of Resolut Adapt® and non-resorbable membranes composed of d-PTFE in the treatment of bone defects associated with periodontitis, and the results showed similarity between the two groups.

ε-policaprolactona (PCL) is also an aliphatic ester that, despite having good biocompatibility, low cost and high mechanical strength, exhibits a resorption time of 2-3 years due to its hydrophobic characteristic, which limits its application in GBR. To get around this inconvenience, PCL has been associated with other aliphatic esters, such as PLA [40,43]. A commercial option for composite membranes from the copolymerization of PCL and PLA is Vivosorb® (Polyganics, The Netherlands), which was originally granted for nerve regeneration and has been used for GBR due to its space-maintaining capacity and low rate of resorption [52]. The effectiveness of this membrane on GBR in mandibular bone defects in rats was compared to e-PTFE and collagen membranes by Gielkens et al. [53]. Vivosorb® demonstrated better bone formation in relation to the control group (untreated) after 12 weeks, however, in relation to collagen membranes and e-PTFE in the same period, they presented less bone formation. This fact is possible due to the reduction in the loss of barrier function of the PCL and PLA membranes after 10 weeks caused by their degradation and consequent resorption.

3.2   Polyurethanes

Polyurethane (PUs) was developed by Otto Bayer and collaborators in 1930. It is a polymer well accepted in the general industry due to its important characteristics, such as: high mechanical resistance, good wear resistance, excellent abrasion and chemical resistance. According to the type of raw material and synthesis methodology, PUs can be part of the composition of various consumer goods (shoes, bedding, upholstery, paints, packaging, biomedical devices, among others), highlighting their versatility [54].

Chemically, it is a polymeric condensation group with a urethane group (—NHCOO—) in its structure. The synthesis is obtained through the polymerization reaction in growth steps between isocyanates (diisocyanate or polyisocyanate) and polyols (low molecular weight oligomeric polyol with terminal hydroxyl groups), in the presence of a suitable catalyst,

Applications of Polymers in Surgery   Materials Research Forum LLC
Materials Research Foundations **123** (2022) 195-215   https://doi.org/10.21741/9781644901892-8

forming linear, branched or crosslinked polymers. Petroleum-derived polyols and isocyanates are widely used and researchers are currently looking for other alternatives to avoid dependence on resources from petrochemicals [54,55].

A natural source of PUs (Poliquil®) well known in the Brazilian flora is Ricinus communis, popularly known as "mamona". This biopolymer with high biocompatibility is extracted from its oil. Osteoconductive and osteointegrated action has already been demonstrated using PUs, which collaborates with the neoformation of bone tissue. Therefore, it is an example of a raw material that can replace dependence on petroleum to develop biomedical devices, for example [56].

It is noteworthy that the synthesis from isocyanates and polyols gives rise to products with weak mechanical properties. Therefore, a chain extender is included in the structure, in order to induce microphase separation between the two thermodynamically incompatible segments. Isocyanate and the chain extender form the rigid segments responsible for the mechanical strength, whereas the polyol is part of the soft segments promoting flexibility to the product [57].

This modification of mechanical properties enables the wide applicability of PUs in various industrial fields. In the biomedical area, we can find it, for example, in the development of Active Pharmaceutical Ingredient (API) release systems, cardiovascular devices (catheters, vascular prostheses, pacemakers, etc.), reconstructive surgery (dressings, breast implants, maxillofacial prostheses, other implants, etc.), in obstetrics and gynecology (condoms, contraceptive sponges, etc.), in textile materials (hospital sheets, surgical drapes, etc.), in dentistry (resins, aligning plate, membrane for GBR) [57,58].

Generally, PU membranes are modified to improve their performance. For example, water-based Pus (dispersion) can replace traditional solvent-based PUs, as they have low viscosity, non-toxic, non-polluting, low cost, applicability and safety. However, depending on the application, sometimes another polymer is needed to improve the viscosity parameter [59].

PUs can also be used in the development of sponges for use in oro-maxillary bone defects and in GBR processes. Composed of polyisocyanate, biocompatible polyester diol, water (foaming agent) and other components, it has a porous core and a dense shell shape, which helps in the osteoconduction and acts as a barrier to prevent the growth of gingival tissue [60].

3.3   Chitosan

Chitosan (CS) is a biopolymer with easily moldable biological, physical and chemical properties; it is biocompatible, bioactive, bacteriostatic, biodegradable and non-toxic.

These characteristics are important for its adaptation to other types of materials and, consequently, its adoption in the development of membranes for GBR and other biomedical devices [61,62].

It was discovered in 1859 by a French researcher called Chales Rouget. After deacetylating the natural polymer chitin with potassium hydroxide, he observed the emergence of a new substance that could be dissolved in acidic solutions. Around 1894, the German researcher Felix Hoppe-Seyler named this material as chitosan. From the early 1990s to the present day, it has been gaining ground in the pharmaceutical industry, biotechnology and agribusiness [63,64].

It is obtained from natural and renewable sources, being the second most abundant biopolymer in nature after cellulose. For industrial production, chitosan is usually extracted from the exoskeleton of crustaceans (deacetylation of chitin; 50% NaOH; 110 °C), which is abundant in coastal strips. It is also found in small amounts in the cell wall of fungi (class Zygomycetes, order Mucorales). In the laboratory, its synthesis can follow two methodologies: 1 - Opening of the ring of the exazoline group of a glycosidic derivative or, 2 - Biosynthesis of glucose itself, which is converted into aminosugars and acetylated, finally, polymerized by enzymatic action [65].

The chemical composition explains its versatility and use in different sectors. It is a linear polysaccharide with active amino groups, which make it reactive for linking new groups through mild reactions, along with the cationic character attributed to the amino groups. As for the structure, it is formed by N-acetyl-D-glucosamine and D-glucosamine units (copolymer) with an amino group (NH_2) and two hydroxyl groups (OH) in each repeated glycosidic unit [66], as shown in. 3.

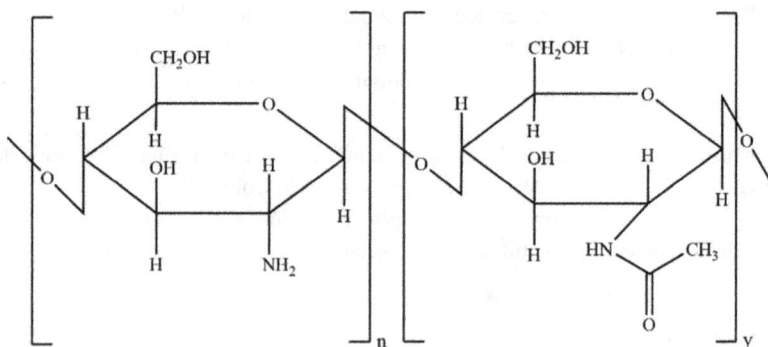

Figure 3.   Chemical structure of chitosan.

Applications of Polymers in Surgery   Materials Research Forum LLC
Materials Research Foundations **123** (2022) 195-215   https://doi.org/10.21741/9781644901892-8

Its physical-chemical, biological characteristics and low cost make CS a biomaterial of choice to compose materials used in dentistry implants and GBR procedures. It can be used in association with other polymers and biomaterials (natural or synthetic), incorporating bioactive molecules, among other strategies that promote osteogenesis [67].

For example, an electrotrophied CS membrane modified with short-chain fatty acids exhibit properties such as: non-swelling, biodegradable and retains the nanofibrous structure in aqueous environments, which is important as it provides increased surface area for the release of API, which can aid fixation, cell growth, exchange of nutrients between bone and epithelial tissues, while remaining occlusive of the cell [68].

GBR membranes containing electrospinned CS submitted to post-spinning processes modified with triethylamine/tert-butyloxycarbonyl (TEA/tBOC) or butyryl-anhydride (BA) to maintain the nanofiber structure, were compared with commercial collagen membrane (BioMend Extend). During the experimental in vivo evaluation, bone growth was observed increasing from 3 to 8 weeks for all tested membranes. However, CS membranes (TEA/tBOC and BA) had more significant results in relation to bone density compared to collagen membrane, showing the capacity of this biomaterial to regenerate bone tissue [69].

In addition to GBR membranes, CS is also an example of a biomaterial used in the composition of scaffolds for bone regeneration in critical bone defects. Studies have already proven its role in the growth of this tissue by increasing the deposition of osteoblasts in mineral-rich matrices. However, it has weak mechanical properties and rapid degradation, a fact that leads to the adoption of other materials to compose these biomedical devices [70,71].

4. Hybrid polymers: a new concept

Hybrid organic-inorganic materials belong to a new class of nanostructured polymers. They combine characteristics of the organic phase and of the inorganic phase and form a product of superior quality when compared to a pure polymer. They can have their structure modified and result in a material with favorable structural, optical and mechanical properties, and even improve the API release profile [72].

Di-ureasil hybrids are an example, they are biocompatible, transparent, flexible and attract special attention because they have better mechanical properties than polyethers [73]. This multifactorial character means that hybrid materials are used in the manufacture of different consumer goods for industry in general (electroanalytical applications, nanofiltration membranes, gas separation, semiconductors, toxic substances adsorbents, biomaterials for osteo-reconstructive surgery, among others) [74,75].

The synthesis of hybrids is carried out through the sol-gel chemical route. During the synthesis of precursors, the solvent normally used is tetrahydrofuran (THF), which is widely used in industry. However, a high exposure to THF can cause some problems in human health and in the environment, which leads researchers to adopt safer solvents (eg: ethanol) [73].

For example, to synthesize hybrid membranes of the ureasyl-polyether type, the sol-gel route is followed as follows: 1 – obtaining the precursors from the reaction of a modified alcooxide 3-isocyanatopropyltriethoxysilane (IsoTrEOS) and modified polymers of polyethylene oxide (NH_2-PEO-NH_2) or based on polypropylene oxide (NH_2-PPO-NH_2); 2 - mixture of alcooxide and modified polymer kept under reflux in absolute ethanol at a temperature of 60 °C for 24 hours; 3 - Formation of the hybrid precursor; 4 – elimination of solvent by heating and reduced pressure (rotary evaporator), obtaining solvent-free precursor; 5 - hydrolysis and condensation reactions promoted by the controlled addition of water, ethanol and an acidic catalyst (HCl) forming a gel; 6 – obtaining the membranes by drying the gel at room temperature in a desiccator for 3 days [76-80].

According to the type of interaction between the organic and inorganic phases, they are classified into two types: Class I - interact by weak bonds (Van der Waals, electrostatic, π-π interaction, hydrogen bond) and Class II - interact by strong bonds (covalent or ionic-covalent bonds). However, class II hybrid materials may have the same type of bond as found in class I hybrids [81,82].

An advantage of the hybrid membranes used in GBR is that, for example, osteogenic growth peptide (OGP) can be added as an osteoinductive substance. In addition to exercising its physical barrier action, it also promotes the formation of bone tissue with the quantity and quality necessary for use in implant dentistry [83].

Another example of hybrid membranes for GBR are biomimetic nanofibers, they have favorable mechanical characteristics, release growth factor in a controlled manner, are antibacterial and demonstrate the regenerative action of bone tissue within expected parameters [84].

Conclusion

Advances in polymer science enable its use in the composition of different biomedical products, including membranes for GBR in implantology. Its various physicochemical, mechanical and biological characteristics are suitable for handling during surgery and act as a physical barrier, which promotes an ideal environment to regenerate bone tissue in a healthy way. For this reason, polymers have been used for over 20 years at GBR and have been constantly improved, such as the new generation of membranes combining polymers

Applications of Polymers in Surgery   Materials Research Forum LLC
Materials Research Foundations **123** (2022) 195-215   https://doi.org/10.21741/9781644901892-8

with active biomolecules to ensure better treatment effectiveness. However, the high cost of these products limits access by patients. To get around this, the low cost and easy access of some polymeric raw materials can move this market. This enables the development of efficient, affordable membranes and helps to reduce the cost of the procedure, which may become more adopted by edentulous patients.

References

[1] A. Gupta, D.A. Felton, T. Jemt, S. Koka, Rehabilitation of edentulism and mortality: A Systematic review, J. Prosthodont. 28 (2019) 526-535. https://doi.org/10.1111/jopr.12792

[2] S. Chowdhury, P.P. Chakraborty, Universal health coverage - There is more to it than meets the eye, J. Fam. Med. Prim. Care. 6 (2017) 169-170. https://doi.org/10.4103/jfmpc.jfmpc_13_17

[3] M. Cardoso, I. Balducci, D.M. Telles, E.J.V. Lourenço, L. Nogueira-Júnior, Edentulismo no Brasil: Tendências, projeções e expectativas até 2040, Cienc. e Saude Coletiva. 21 (2016) 1239-1246. https://doi.org/10.1590/1413-81232015214.13672015

[4] M. Khan, J. Ullah, H. Qazi, Frenquency of common factors responsible for non-treatment of partial edentulism, Pakistan Oral Dent. J. 38 (2018) 9-13.

[5] M. Gul, M. Yucesan, F. Serin, E. Celik, A simulation model to improve production processes in an implant manufacturing plant, 21st Int. Res. Conf. "Trends Dev. Mach. Assoc. Technol. (2018) 177-180.

[6] Z. Özkurt-Kayahan, C. Özçakır-Tomruk, E. Kazazoğlu, Partial edentulism and treatment options, Yeditepe Dent. J. 13 (2017) 31-36. https://doi.org/10.5505/yeditepe.2017.62207

[7] K.V.N. Avinash, R. Sampathkumar, V. Raghu, C. Reddy, S. Munireddy, J. Shetty, Implants in narrow ridge situations: A piezo-surgical approach for alveolar ridge split, Int. J. Appl. Dent. Sci. 5 (2019) 244-248.

[8] J.A. Oshiro, M.R. Sato, C.R. Scardueli, G.J.P.L. de Oliveira, M.P. Abucafy, M. Chorilli, Bioactive molecule-loaded drug delivery systems to optimize bone tissue repair, Curr. Protein Pept. Sci. 18 (2017) 1-14. https://doi.org/10.2174/1389203718666170328111605

[9] Y. Zhang, X. Zhang, B. Shi, R. Miron, Membranes for guided tissue and bone regeneration, Ann. Oral Maxillofac. Surg. 1 (2013) 1-10. https://doi.org/10.13172/2052-7837-1-1-451

[10] Y.K. Kim, J.K. Ku, Guided bone regeneration, J. Korean Assoc. Oral Maxillofac. Surg. 46 (2020) 361-366. https://doi.org/10.5125/jkaoms.2020.46.5.361

[11] M. Toledano, J.L. Gutierrez-Pérez, A. Gutierrez-Corrales, M.A. Serrera-Figallo, M. Toledano-Osorio, J.I. Rosales-Leal, M. Aguilar, R. Osorio, D. Torres-Lagares, Novel non-resorbable polymeric-nanostructured scaffolds for guided bone regeneration, Clin. Oral Investig. 24 (2020) 2037-2049. https://doi.org/10.1007/s00784-019-03068-8

[12] S.W. Lee, S.G. Kim, Membranes for the Guided Bone Regeneration, Maxillofac. Plast. Reconstr. Surg. 36 (2014) 239-246. https://doi.org/10.14402/jkamprs.2014.36.6.239

[13] G. Nechifor, E.E. Totu, A.C. Nechifor, I. Isildak, O. Oprea, C.M. Cristache, Non-resorbable nanocomposite membranes for guided bone regeneration based on polysulfone-quartz fiber grafted with nano-TiO2, Nanomaterials. 9 (2019) 1-22. https://doi.org/10.3390/nano9070985

[14] J.Y. Park, J.H. Lee, C.H. Kim, Y.J. Kim, Fabrication of polytetrafluoroethylene nanofibrous membranes for guided bone regeneration, RSC Adv. 8 (2018) 34359-34369. https://doi.org/10.1039/C8RA05637D

[15] L.W. McKeen, Plastics Used in Medical Devices, in: Handb. Polym. Appl. Med. Med. Devices, 2014: p21-53. https://doi.org/10.1016/B978-0-323-22805-3.00003-7

[16] L.W. McKeen, Fluoropolymers, in: Film Prop. Plast. Elastomers, 2012: pp. 255-313. https://doi.org/10.1016/B978-1-4557-2551-9.00011-6

[17] C. Dahlin, A. Linde, J. Gottlow, S. Nyman, Healing of bone defects by guided tissue regeneration, Plast. Reconstr. Surg. 81 (1988) 672-676. https://doi.org/10.1097/00006534-198805000-00004

[18] Y.D. Rakhmatia, Y. Ayukawa, A. Furuhashi, K. Koyano, Current barrier membranes: Titanium mesh and other membranes for guided bone regeneration in dental applications, J. Prosthodont. Res. 57 (2013) 3-14. https://doi.org/10.1016/j.jpor.2012.12.001

[19] A. Ranjbarzadeh-Dibazar, J. Barzin, P. Shokrollahi, Microstructure crystalline domains disorder critically controls formation of nano-porous/long fibrillar morphology of ePTFE membranes, Polymer (Guildf). 121 (2017) 75-87. https://doi.org/10.1016/j.polymer.2017.06.003

[20] M. Trobos, A. Juhlin, F.A. Shah, M. Hoffman, H. Sahlin, C. Dahlin, In vitro evaluation of barrier function against oral bacteria of dense and expanded

polytetrafluoroethylene (PTFE) membranes for guided bone regeneration, Clin. Implant Dent. Relat. Res. 20 (2018) 738-748. https://doi.org/10.1111/cid.12629

[21] R. Dimitriou, G.I. Mataliotakis, G.M. Calori, P. V. Giannoudis, The role of barrier membranes for guided bone regeneration and restoration of large bone defects: Current experimental and clinical evidence, BMC Med. 10 (2012). https://doi.org/10.1186/1741-7015-10-81

[22] J.M. Carbonell, I.S. Martín, A. Santos, A. Pujol, J.D. Sanz-Moliner, J. Nart, High-density polytetrafluoroethylene membranes in guided bone and tissue regeneration procedures: A literature review, Int. J. Oral Maxillofac. Surg. 43 (2014) 75-84. https://doi.org/10.1016/j.ijom.2013.05.017

[23] A. Cucchi, E. Vignudelli, A. Napolitano, C. Marchetti, G. Corinaldesi, Evaluation of complication rates and vertical bone gain after guided bone regeneration with non-resorbable membranes versus titanium meshes and resorbable membranes. A randomized clinical trial, Clin. Implant Dent. Relat. Res. 19 (2017) 821-832. https://doi.org/10.1111/cid.12520

[24] J.I. Sasaki, G.L. Abe, A. Li, P. Thongthai, R. Tsuboi, T. Kohno, S. Imazato, Barrier membranes for tissue regeneration in dentistry, Biomater. Investig. Dent. 8 (2021) 54-63. https://doi.org/10.1080/26415275.2021.1925556

[25] I.A. Urban, M.H.A. Saleh, A. Ravidà, A. Forster, H.L. Wang, Z. Barath, Vertical bone augmentation utilizing a titanium-reinforced PTFE mesh: A multi-variate analysis of influencing factors, Clin. Oral Implants Res. (2021) 1-12. https://doi.org/10.1111/clr.13755

[26] A. Cucchi, P. Ghensi, Vertical Guided Bone Regeneration using Titanium-reinforced d-PTFE Membrane and prehydrated corticocancellous Bone Graft, Open Dent. J. 8 (2014) 194-200. https://doi.org/10.2174/1874210601408010194

[27] P. Windisch, K. Orban, G.E. Salvi, A. Sculean, B. Molnar, Vertical-guided bone regeneration with a titanium-reinforced d-PTFE membrane utilizing a novel split-thickness flap design: a prospective case series, Clin. Oral Investig. 25 (2021) 2969-2980. https://doi.org/10.1007/s00784-020-03617-6

[28] L. Zoran, B. Marija, M. Radomir, M. Stevo, M. Marko, D. Igor, Comparison of resorbable membranes for guided bone regeneration of human and bovine origin, Acta Vet. Brno. 64 (2014) 477-492. https://doi.org/10.2478/acve-2014-0045

[29] N.K. Soldatos, P. Stylianou, P. Koidou, N. Angelov, R. Yukna, G.E. Romanos, Limitations and options using resorbable versus nonresorbable membranes for

successful guided bone regeneration, Quintessence Int. (Hanover Park. IL). 48 (2017) 131-147.

[30] L.P. Mau, C.W. Cheng, P.Y. Hsieh, A.A. Jones, Biological complication in guided bone regeneration with a polylactic acid membrane: A case report, Implant Dent. 21 (2012) 171-174. https://doi.org/10.1097/ID.0b013e31824eece1

[31] C. Fidalgo, M.A. Rodrigues, T. Peixoto, J. V. Lobato, J.D. Santos, M.A. Lopes, Development of asymmetric resorbable membranes for guided bone and surrounding tissue regeneration, J. Biomed. Mater. Res. Part A. 106 (2018) 2141-2150. https://doi.org/10.1002/jbm.a.36420

[32] P. Aprile, D. Letourneur, T. Simon-Yarza, Membranes for guided bone regeneration: A Road from bench to bedside, Adv. Healthc. Mater. 9 (2020) 2000707. https://doi.org/10.1002/adhm.202000707

[33] E.K. Efthimiadou, M. Theodosiou, G. Toniolo, N.Y. Abu-Thabit, Stimuli-responsive biopolymer nanocarriers for drug delivery applications, in: Stimuli Responsive Polym. Nanocarriers Drug Deliv. Appl., 2018: pp. 405-432. https://doi.org/10.1016/B978-0-08-101997-9.00019-9

[34] H. Zhang, Q. Zhang, Q. Cai, Q. Luo, X. Li, X. Li, In-reactor engineering of bioactive aliphatic polyesters via magnesium-catalyzed polycondensation for guided tissue regeneration, Chem. Eng. J. 424 (2021) 130432. https://doi.org/10.1016/j.cej.2021.130432

[35] S.J. Huang, Biodegradation, in: Compr. Polym. Sci. Suppl., 1989: pp. 597-606. https://doi.org/10.1016/B978-0-08-096701-1.00201-9

[36] Y. Qin, A brief description of textile fibers, in: Med. Text. Mater., 2016: pp. 23-42. https://doi.org/10.1016/B978-0-08-100618-4.00003-0

[37] L.M. Caballero, S.M. Silva, S.E. Moulton, Growth factor delivery: Defining the next generation platforms for tissue engineering, J. Control. Release. 306 (2019) 40-58. https://doi.org/10.1016/j.jconrel.2019.05.028

[38] M. Chi, M. Qi, A. Lan, P. Wang, M.D. Weir, M.A. Melo, X. Sun, B. Dong, C. Li, J. Wu, L. Wang, H.H.K. Xu, Novel bioactive and therapeutic dental polymeric materials to inhibit periodontal pathogens and biofilms, Int. J. Mol. Sci. 20 (2019) 278. https://doi.org/10.3390/ijms20020278

[39] A. Zamboulis, E.A. Nakiou, E. Christodoulou, D.N. Bikiaris, Polyglycerol hyperbranched polyesters: Synthesis, properties and pharmaceutical and biomedical applications, Int. J. Mol. Sci. 20 (2019) 278. https://doi.org/10.3390/ijms20246210

[40] P. Gentile, V. Chiono, C. Tonda-turo, A.M. Ferreira, G. Ciardelli, Polymeric membranes for guided bone regeneration, Biotechnol. J. 6 (2011) 1187-1197. https://doi.org/10.1002/biot.201100294

[41] T. Casalini, F. Rossi, A. Castrovinci, G. Perale, A Perspective on polylactic acid-based polymers use for nanoparticles synthesis and applications, Front. Bioeng. Biotechnol. 7 (2019) 1-16. https://doi.org/10.3389/fbioe.2019.00001

[42] K. Asano, T. Matsuno, Y. Tabata, T. Satoh, Preparation of thermoplastic poly(L-lactic acid) membranes for guided bone regeneration.pdf, Int. J. Oral Maxillofac. Implants. 28 (2013) 973-981. https://doi.org/10.11607/jomi.2729

[43] J. Wang, L. Wang, Z. Zhou, H. Lai, P. Xu, L. Liao, J. Wei, Biodegradable polymer membranes applied in guided bone / tissue regeneration: A review, Polymers (Basel). 8 (2016) 115. https://doi.org/10.3390/polym8040115

[44] D. Kim, T. Kang, D. Gober, C. Orlich, A Liquid membrane as a barrier membrane for guided bone regeneration, ISRN Dent. 2011 (2011) 468282. https://doi.org/10.5402/2011/468282

[45] D. Schneider, F.E. Weber, C. Andreoni, R.E. Jung, D. Schneider, A randomized controlled clinical multicenter trial comparing the clinical and histological performance of a new, modified polylactide-co-glycolide acid membrane to an expanded polytetrafluorethylene membrane in guided bone regeneration procedures, Clin. Oral Implants Res. 25 (2013) 150-158. https://doi.org/10.1111/clr.12132

[46] D. Gorth, T.J. Webster, Matrices for tissue engineering and regenerative medicine, in: Biomater. Artif. Organs, 2011: pp. 270-286. https://doi.org/10.1533/9780857090843.2.270

[47] O.S. Manoukian, N. Sardashti, C. Health, U. States, Biomaterials for tissue engineering and regenerative medicine, in: Encycl. Biomed. Eng., 2019: pp. 462-482. https://doi.org/10.1016/B978-0-12-801238-3.64098-9

[48] E.M. Prieto, S.A. Guelcher, Tailoring properties of polymeric biomedical foams, in: Biomed. Foam. Tissue Eng. Appl., 2014: pp. 129-162. https://doi.org/10.1533/9780857097033.1.129

[49] Y.D. Rakhmatia, Y. Ayujawa, A. Furuhashi, K. Koyano, Current barrier membranes : Titanium mesh and other membranes for guided bone regeneration in dental applications, J. Prosthodont. Res. 57 (2013) 3-14. https://doi.org/10.1016/j.jpor.2012.12.001

[50] E. Milella, P.A. Ramires, E. Brescia, G. La Sala, L. Di Paola, V. Bruno, B. Unit, S.S. Appia, Physicochemical, mechanical, and biological properties of commercial membranes for GTR, J. Biomed. Mater. Res. 58 (2001) 427-435. https://doi.org/10.1002/jbm.1038

[51] A. Wadhawan, T. Gowda, D. Mehta, Gore-tex® versus resolut adapt® GTR membranes with perioglas® in periodontal regeneration, Contemp. Clin. Dent. 3 (2012) 406. https://doi.org/10.4103/0976-237X.107427

[52] I. Elgali, O. Omar, C. Dahlin, P. Thomsen, Guided bone regeneration: materials and biological mechanisms revisited, Eur. J. Oral Sci. 125 (2017) 315-337. https://doi.org/10.1111/eos.12364

[53] P.F.M. Gielkens, J. Schortinghuis, J.R. De Jong, G.M. Raghoebar, B. Stegenga, R.R.M. Bos, Vivosorb®, Bio-Gide®, and Gore-Tex ® as barrier membranes in rat mandibular defects: An evaluation by microradiography and micro-CT, Clin. Oral Implants Res. 19 (2008) 516-521. https://doi.org/10.1111/j.1600-0501.2007.01511.x

[54] M. Ghasemlou, F. Daver, E.P. Ivanova, B. Adhikari, Polyurethanes from seed oil-based polyols: A review of synthesis, mechanical and thermal properties, Ind. Crops Prod. 142 (2019) 111841. https://doi.org/10.1016/j.indcrop.2019.111841

[55] M.A. Sawpan, Polyurethanes from vegetable oils and applications: a review, J. Polym. Res. 25 (2018) 1-15. https://doi.org/10.1007/s10965-017-1408-z

[56] T.P.T. De Sousa, M.S.T. Da Costa, R. Guilherme, W. Orcini, L. de A. Holgado, E.M. Varize Silveira, O. Tavano, A.G. Magdalena, S.A. Catanzaro-Guimarães, A. Kinoshita, Polyurethane derived from Ricinus Communis as graft for bone defect treatments, Polimeros. 28 (2018) 246-255. https://doi.org/10.1590/0104-1428.03617

[57] M.B. Lowinger, S.E. Barrett, F. Zhang, R.O. Williams, Sustained release drug delivery applications of polyurethanes, Pharmaceutics. 10 (2018) 1-19. https://doi.org/10.3390/pharmaceutics10020055

[58] T. Pivec, M. Sfiligoj-Smole, P. Gašparič, K.S. Kleinschek, Polyurethanes for medical use, Tekstilec. 60 (2017) 182-197. https://doi.org/10.14502/Tekstilec2017.60.182-197

[59] C. Zhang, J. Wang, Y. Xie, L. Wang, L. Yang, J. Yu, A. Miyamoto, F. Sun, Development of FGF-2-loaded electrospun waterborne polyurethane fibrous membranes for bone regeneration, Regen. Biomater. 8 (2021) 1-9. https://doi.org/10.1093/rb/rbaa046

[60] S.M. Giannitelli, F. Basoli, P. Mozetic, P. Piva, F.N. Bartuli, F. Luciani, C. Arcuri, M. Trombetta, A. Rainer, S. Licoccia, Graded porous polyurethane foam: A potential scaffold for oro-maxillary bone regeneration, Mater. Sci. Eng. C. 51 (2015) 329-335. https://doi.org/10.1016/j.msec.2015.03.002

[61] G.S. Andrade, D. de B. Andrade, G.G. de Lima, F.F. Padilha, P.A.L. Lima, Prospecção tecnológica de quitosana, fibroína e goma xantana como biomateriais aplicáveis em scaffolds-3d technological forecasting of chitosan, silk fibroin and xanthan gum as biomaterials for scaffolds-3d, GEINTEC. 10 (2020) 5279-5288. https://doi.org/10.7198/geintec.v10i1.1173

[62] M.J.L. Dantas, T.B. Fidéles, R.G. Carrodeguas, M.V.L. Fook, Obtenção e caracterização de esferas de quitosana / hidroxiapatita gerada in situ, Rev. Eletrônica Mater. e Process. 11 (2016) 18-24.

[63] R.A.I. Reshad, T.A. Jishan, N.N. Chowdhury, Chitosan and its broad applications: A brief review, SSRN. 4 (2021) 1-55. https://doi.org/10.2139/ssrn.3842055

[64] B. Tian, Y. Liu, Chitosan-based biomaterials: From discovery to food application, Polym. Adv. Technol. 31 (2020) 2408-2421. https://doi.org/10.1002/pat.5010

[65] T.M.S. Arnaud, T.C.M. Stamford, T.L.M. Stamford, N.P. Stamford, Produção, propriedades e aplicações da quitosana na agricultura e em alimentos, Biotecnol. Apl. à Agro&Indústria. 4 (2017) 503-528. https://doi.org/10.5151/9788521211150-14

[66] E.P. Pinto, W.D.S. Tavares, R.S. Matos, A.M. Ferreira, R.P. Menezes, M.E.H.M. Da Costa, T.M. De Souza, I.M. Ferreira, F.F.O. De Sousa, R.R.M. Zamora, Influence of low and high glycerol concentrations on wettability and flexibility of chitosan biofilms, Quim. Nova. 41 (2018) 1109-1116. https://doi.org/10.21577/0100-4042.20170287

[67] A. Aguilar, N. Zein, E. Harmouch, B. Hafdi, F. Bornert, O. Damien, F. Clauss, F. Fioretti, O. Huck, N. Benkirane-jessel, G. Hua, Biodental Engineering V, Biodental Eng. V. (2019) 1-17.

[68] V.P. Murali, F.D. Guerra, N. Ghadri, J.M. Christian, S.H. Stein, J.A. Jennings, R.A. Smith, J.D. Bumgardner, Simvastatin loaded chitosan guided bone regeneration membranes stimulate bone healing, J. Periodontal Res. (2021) 1-8. https://doi.org/10.1111/jre.12883

[69] H. Su, T. Fujiwara, K.M. Anderson, A. Karydis, M.N. Ghadri, J.D. Bumgardner, A comparison of two types of electrospun chitosan membranes and a collagen membrane in vivo, Dent. Mater. 37 (2021) 60-70. https://doi.org/10.1016/j.dental.2020.10.011

[70] R. Ikono, N. Li, N.H. Pratama, A. Vibriani, D.R. Yuniarni, M. Luthfansyah, B.M. Bachtiar, E.W. Bachtiar, K. Mulia, M. Nasikin, H. Kagami, X. Li, E. Mardliyati, N.T. Rochman, T. Nagamura-Inoue, A. Tojo, Enhanced bone regeneration capability of chitosan sponge coated with TiO2 nanoparticles, Biotechnol. Reports. 24 (2019). https://doi.org/10.1016/j.btre.2019.e00350

[71] A.K. Mahanta, D.K. Patel, P. Maiti, Nanohybrid scaffold of chitosan and functionalized graphene oxide for controlled drug delivery and bone regeneration, ACS Biomater. Sci. Eng. 5 (2019) 5139-5149. https://doi.org/10.1021/acsbiomaterials.9b00829

[72] J.A. Oshiro Junior, L.M. Shiota, L.A. Chiavacci, Desenvolvimento de formadores de filmes poliméricos orgânico-inorgânico para liberação controlada de fármacos e tratamento de feridas, Rev. Mater. 19 (2014) 24-32. https://doi.org/10.1590/S1517-70762014000100005

[73] J.F. Mendes, J. Augusto, O. Junior, C. Garcia, L.A. Chiavacci, Synthesis of ureasil-polyether film forming materials by using environmentally friendly solvent, (2021) 1-10. https://doi.org/10.4322/2179-443X.0730

[74] J.M. García-Martínez, E.P. Collar, Organic-Inorganic Hybrid Materials by ALD, Polymers (Basel). 13 (2020) 4. https://doi.org/10.3390/polym13010086

[75] B. Samiey, C.H. Cheng, J. Wu, Organic-inorganic hybrid polymers as adsorbents for removal of heavy metal ions from solutions: A review, Materials (Basel). 7 (2014) 673-726. https://doi.org/10.3390/ma7020673

[76] J.A. Oshiro Junior, G.R. Mortari, R.M. de Freitas, E. Marcantonio-Junior, L. Lopes, L.C. Spolidorio, R.A. Marcantonio, L.A. Chiavacci, Assessment of biocompatibility of ureasil-polyether hybrid membranes for future use in implantodontology, Int. J. Polym. Mater. Polym. Biomater. 65 (2016) 647-652. https://doi.org/10.1080/00914037.2016.1157796

[77] J.A. Oshiro Junior, A. Lusuardi, L.A. Chiavacci, M.T. Cuberes, Nanostructural arrangements and surface morphology on ureasil-polyether films loaded with dexamethasone acetate, Nanomaterials. 11 (2021) 1-19. https://doi.org/10.3390/nano11061362

[78] K.M.N. Costa, M.R. Sato, T.L.A. Barbosa, M.G.F. Rodrigues, A.C.D. Medeiros, B.P.G.L. Damasceno, J.A. Oshiro-Junior, Curcumin-loaded micelles dispersed in ureasil-polyether materials for a novel sustained-release formulation, Pharmaceutics. 13 (2021) 1-10. https://doi.org/10.3390/pharmaceutics13050675

[79] J.A. Oshiro Junior, N.J. Nasser, B.G. Chiari-Andréo, M.T. Cuberes, L.A. Chiavacci, Study of triamcinolone release and mucoadhesive properties of macroporous hybrid films for oral disease treatment, Biomedical Physics & Engineering Express. 4 (2018) 1-9. https://doi.org/10.1088/2057-1976/aaa84b

[80] J.A. Oshiro Junior, C.R. Scardueli, G.P.L. de Oliveira, R.A.C. Marcantonio, L.A. Chiavacci, Development of ureasil-polyether membranes for guided bone regeneration, Biomedical Physics & Engineering Express. 3 (2017) 1-10. https://doi.org/10.1088/2057-1976/aa56a6

[81] M.M. Sanagi, N.T. Ng, F.M. Marsin, M.R. Zinalibdin, Z.A. Sutirman, A.S.A. Keyon, W.A.W. Ibrahim, Facile organic-inorganic hybrid sorbents for extraction of pollutants from aqueous samples - A review, Malaysian J. Anal. Sci. 23 (2019) 561-571.

[82] J.A. Oshiro-Junior, M.P. Abuçafy, E.B. Manaia, B.L da Silva, B.G. Chiari-Andréo, L.A. Chiavacci, Drug delivery systems obtained from silica based organic-inorganic hybrids, Polymers. 8 (2016) 1-14. https://doi.org/10.3390/polym8040091

[83] J.A. Oshiro-Junior, R.M. Barros, C.G. da Silva, C.C. de Souza, C.R. Scardueli, C.C. Marcantonio, P.R. da Silva Saches, L. Mendes, E.M. Cilli, R.A.C. Marcantonio, L.A. Chiavacci, In vivo effectiveness of hybrid membranes with osteogenic growth peptide for bone regeneration, J. Tissue Eng. Regen. Med. (2021) 1-10. https://doi.org/10.1002/term.3226

[84] Q. Yuan, L. Li, Y. Peng, A. Zhuang, W. Wei, D. Zhang, Y. Pang, X. Bi, Biomimetic nanofibrous hybrid hydrogel membranes with sustained growth factor release for guided bone regeneration, Biomater. Sci. 9 (2021) 1256-1271. https://doi.org/10.1039/D0BM01821J

Keyword Index

About the Editors

Dr. Inamuddin is working as Assistant Professor at the Department of Applied Chemistry, Aligarh Muslim University, Aligarh, India. He obtained Master of Science degree in Organic Chemistry from Chaudhary Charan Singh (CCS) University, Meerut, India, in 2002. He received his Master of Philosophy and Doctor of Philosophy degrees in Applied Chemistry from Aligarh Muslim University (AMU), India, in 2004 and 2007, respectively. He has extensive research experience in multidisciplinary fields of Analytical Chemistry, Materials Chemistry, and Electrochemistry and, more specifically, Renewable Energy and Environment. He has worked on different research projects as project fellow and senior research fellow funded by University Grants Commission (UGC), Government of India, and Council of Scientific and Industrial Research (CSIR), Government of India. He has received Fast Track Young Scientist Award from the Department of Science and Technology, India, to work in the area of bending actuators and artificial muscles. He has also received the Sir Syed Young Researcher of the Year Award 2020 from Aligarh Muslim University. He has completed four major research projects sanctioned by University Grant Commission, Department of Science and Technology, Council of Scientific and Industrial Research, and Council of Science and Technology, India. He has published 199 research articles in international journals of repute and nineteen book chapters in knowledge-based book editions published by renowned international publishers. He has published 150 edited books with Springer (U.K.), Elsevier, Nova Science Publishers, Inc. (U.S.A.), CRC Press Taylor & Francis Asia Pacific, Trans Tech Publications Ltd. (Switzerland), IntechOpen Limited (U.K.), Wiley-Scrivener, (U.S.A.) and Materials Research Forum LLC (U.S.A). He is a member of various journals' editorial boards. He is also serving as Associate Editor for journals (Environmental Chemistry Letter, Applied Water Science and Euro-Mediterranean Journal for Environmental Integration, Springer-Nature), Frontiers Section Editor (Current Analytical Chemistry, Bentham Science Publishers), Editorial Board Member (Scientific Reports-Nature), Editor (Eurasian Journal of Analytical Chemistry), and Review Editor (Frontiers in Chemistry, Frontiers, U.K.) He is also guest-editing various special thematic special issues to the journals of Elsevier, Bentham Science Publishers, and John Wiley & Sons, Inc. He has attended as well as chaired sessions in various international and national conferences. He has worked as a Postdoctoral Fellow, leading a research team at the Creative Research Initiative Center for Bio-Artificial Muscle, Hanyang University, South Korea, in the field of renewable energy, especially biofuel cells. He has also worked as a Postdoctoral Fellow at the Center of Research Excellence in Renewable Energy, King Fahd University of Petroleum and Minerals, Saudi Arabia, in the field of polymer electrolyte membrane fuel cells and computational fluid dynamics of polymer electrolyte membrane fuel cells. He is a life member of the Journal of the Indian Chemical Society. His research interest includes ion exchange materials, a sensor for heavy metal ions, biofuel cells, supercapacitors and bending actuators.

Dr. Tariq Altalhi, joined the Department of Chemistry at Taif University, Saudi Arabia as Assistant Professor in 2014. He received his doctorate degree from University of Adelaide, Australia in the year 2014 with Dean's Commendation for Doctoral Thesis Excellence. He was promoted to the position of the head of Chemistry Department at Taif university in 2017 and Vice Dean of Science college in 2019 till now. In 2015, one of his works was nominated for Green Tech awards from Germany, Europe's largest environmental and business prize, amongst top 10 entries. He has co-edited various scientific books. His group is involved in fundamental multidisciplinary research in nanomaterial synthesis and engineering, characterization, and their application in molecular separation, desalination, membrane systems, drug delivery, and biosensing. In addition, he has established key contacts with major industries in Kingdom of Saudi Arabia.

Dr. Mohammad Luqman has 12+ years of post-PhD experience in Teaching, Research, and Administration. Currently, he is serving as an Assistant Professor of Chemical Engineering in Taibah University, Saudi Arabia. Before joining here, he served as an Assistant Professor in College of Applied Science at A'Sharqiyah University, Oman, and in College of Engineering at King Saud University, Saudi Arabia. He served as a Research Engineer in SAMSUNG Cheil Industries, South Korea. Moreover, he served as a post-doctoral fellow at Artificial Muscle Research Center, Konkuk University, South Korea, in the field of Ionic Polymer Metal Composites for the development of Artificial Muscles, Robotic Actuators and Dynamic Sensors. He earned his PhD degree in the field of Ionomers (Ion-containing Polymers), from Chosun University, South Korea. He successfully served as an Editor to three books, published by world renowned publishers. He published numerous high-quality papers, and book chapters. He is serving as an Editor and editorial/review board members to many International SCI and Non-SCI journals. He has attracted a few important research grants from industry and academia. His research interests include but not limited to Development of Ionomer/Polyelectrolyte/non-ionic Polymer Nanocomposites/Blends for Smart and Industrial/Engineering Applications.

Dr. Hamida-Tun-Nisa Chisti is currently working as an Associate Professor in the Department of Chemistry, National Institute of Technology Srinagar, India. She has received her B.Sc. Hons., Masters and Ph.D. in Chemistry from Aligarh Muslim University, Aligarh, and later has joined her services as Lecturer in 2008 at NIT Srinagar. Her research focus is in the fields of Inorganic Chemistry, Environmental

Chemistry, and Material Chemistry. She has published several research articles in international journals of repute. She has authored 4 books and 6 book chapters. She is a Member Royal Society of Chemistry (MRSC), ACS and also a life member of Asian Polymer Association, Indian Council of Chemists, Life member of the International Association of Engineers (**IAENG**), Hong Kong, Life member of Eco Ethics International Union (**EEIU**), Germany and many more. She is also serving as a reviewer for many journals.